Cybersecurity and Identity Access Management

Bharat S. Rawal · Gunasekaran Manogaran ·
Alexender Peter

Cybersecurity and Identity Access Management

 Springer

Bharat S. Rawal
Department of Computer and Data Science
Capitol Technology University
Laurel, MD, USA

Gunasekaran Manogaran
Howard University
Washington, DC, USA

Alexender Peter
Blockchain Datacenter Inc.
Virginia, VA, USA

ISBN 978-981-19-2660-0   ISBN 978-981-19-2658-7   (eBook)
https://doi.org/10.1007/978-981-19-2658-7

This Springer imprint is published by the registered company Springer Nature Singapore Pte Ltd.
The registered company address is: 152 Beach Road, #21-01/04 Gateway East, Singapore 189721,
Singapore

Preface

Cybersecurity protects the computer systems' security and their hardware network from theft or damages both the program and the electronic data. The services they provide are distracting or misdirected. As the area increases, it is increasingly becoming an essential component of computer systems. The growth of the Internet, wireless networks like Bluetooth and Wi-Fi, and connected "intelligent" devices, smartphones, TVs, and various "Internet of Something" devices, puts more demand for security. The practice of protecting systems is cybersecurity, virtual assault networks, and services. The usual purpose of these cyberattacks is to access and modify or lose sensitive data. Cyber technology is critical for organizations and the public who need information security tools to guard against cyber threats. It is vital to protect three key entities: machines, mobile devices, endpoint devices, networks, routers, and cloud. Standard technology for protecting these organizations includes firewalls for next-generation, filtering DNS, protection of malware, antivirus, and email security solutions. The challenges in cybersecurity are increasing with the evolution of malware, ransomware, AI growth, IoT warning, revolution blockchain, and the increasing use of serverless apps.

This book covers all the challenges mentioned above. The malware's evolution may worsen than a propagating virus; this locks the customer's information systems. The business can easily go out of business if you cannot meet the egregious demands of the cybercriminal. Attacks in ransomware are one of the most rapidly growing areas of cybercrime. The major problem with cyberattacks is recovery, and safety professionals seldom can deal immediately with them. Since there is no need to sleep with artificial intelligence, they can oppose defense systems when downloading malware. A risk to prepare security professionals to meet login convergence requirements, user verification, time-out, authentication of two factors, and other advanced defense protocols. The use of blockchain requires additional security integrations with the traditional cybersecurity approach. Serverless apps can call up cyberattacks.

Chapter 1 explains the cybersecurity introduction, fundamentals, principles, importance, challenges, implementation, cyberattack, and types.

Chapter 2 explains the hacking introduction, classification, ethical hacking, cybercrime, penetration test, and a security assessment phase.

Chapter 3 explains the security hacking, techniques, and tools for ethical hacking, security testing, physical security, risk assessment, and protection techniques.

Chapter 4 explains the networking, topology types, operating system, hardware networking, planning and design of the network, problem-solving, and network security.

Chapter 5 explains the cyberattack, vulnerability, network security governance framework, risk assessment, security evaluation and protection, cyber danger, and supply chain.

Chapter 6 explains the malware, vulnerability, network security governance framework, risk assessment, security evaluation and protection, cyber danger, and supply chain.

Chapter 7 explains the firewalls, vulnerability, network security governance framework, risk assessment, security evaluation and protection, cyber danger, and supply chain.

Chapter 8 explains the cryptography, vulnerability, network security governance framework, risk assessment, security evaluation and protection, cyber danger, and supply chain.

Chapter 9 explains the control of physical and logical access to assets.

Chapter 10 explains the management of the identification and authentication of people, devices, and services.

Chapter 11 explains the integration of identity as a third-party service.

Chapter 12 explains the implementation and management of authorization mechanisms.

Chapter 13 explains the management of the identity and access provisioning life cycle.

Chapter 14 explains the conduct of security control testing.

Chapter 15 explains the collection of security process data.

Chapter 16 explains the recovery strategies for databases.

Chapter 17 explains the analysis of test output and generation of a report.

Chapter 18 explains the ensured appropriate asset retention.

Chapter 19 explains the determined information and security controls.

Laurel, USA   Bharat S. Rawal
Washington, USA   Gunasekaran Manogaran
Virginia, USA   Alexender Peter

Contents

Chapter 1
Cybersecurity for Beginners

1.1 Introduction to Cybersecurity

New possibilities and future growth tools are developing across all organizations' sizes in an era of technical innovation. Still, these latest technologies have often introduced unique challenges to the world's economy and population. The safety and security of companies must be ensured by steps taken to maintain protection. Data and information hacking in organizations seem to be becoming a practice. Therefore, the characteristics of information security must be recognized. Data and information hacking in organizations have almost become a practice. The features of information protection will be identified. It is a big concern for any organization. Cybersecurity technology is seen as a new way for investors and ordinary citizens to reach a wide variety of resources and resources to easily, economically, and effectively carry out their activities. This leads fraudsters to implement fraudulent systems in parallel. Internet media is a valuable tool for serious crime growth. As the internet overgrows, online hackers are attempting to create fraudulent plans in several ways.

In general, cybersecurity is connected to the internet. For several years, experts and politicians have become more concerned with protecting ICT systems from cyberattacks than unauthorized persons have been attempting intentionally to obtain exposure to ICT systems to reach the aims of attack, perturbation, disruption, and other illegal acts. Cybersecurity includes finding, evaluating, and minimizing vulnerabilities and decreasing trust in "new" IT institutions and resources due to the globalization of the supply chains, growing exponentially complicated devices and computer code, rising accessible digital networks, and accidental and strategic exploitations and barriers by human and institutional actions. Over the next two years, several professionals anticipate the number and intensity of cyberattacks.

In addition, cybersecurity is related to technological terms, digital security, which is defined in Federal law as securing information and information systems to provide privacy, Confidentiality, and availability against unauthorized entry, usage, disclosure, interruption, modification, or damage. Integrity requires protection from unauthorized change and loss of information, which allows for non-repudiation and authenticity. Confidentiality involves the protection of authorized access and disclosure constraints, including ways to protect data and safety. Availability means that connectivity and utilization are timely and secure. Cybersecurity focuses on protecting unauthorized or illegal entry, modification, or damage to devices, networks, programs, and data. Governments, the state, business corporations, financial agencies, hospitals, and other businesses collect processes and process large numbers of confidential details on devices. Given the increasing amount and sophistication of cyberattacks, greater focus is required to secure confidential place, personal, and national security information.

1.2 Necessity of Cybersecurity

Most organizations, corporate, government, financial, military, and medical organizations, continuously collect large volumes of data to store and process their applications. At the time of this process, their data should be protected from third-party access; thus, cybersecurity is more important. Once the organizations are failing to protect their data, then they are facing negative consequences. Therefore, sensitive information must be saved by using the cybersecurity concept before transmitting data via the network. The techniques' growth leads to several cyberattacks that steal the organizations and institution's sensitive information entirely. Hence, cybersecurity must be implemented to manage sensitive information.

1.3 Cybersecurity Challenges

There are emerging cyberspace threats such as mobile phones that raise safety issues. Smartphone and cloud networking imply that users face a whole different range of interconnectivity concerns involving new legislation and creative technology. In countries worldwide that use information and communication technologies (ICT), the cybersecurity challenge is standard, as they use specific applications and hardware. The most important reasons are that both countries use TCP/IPs as communication protocols, based on one program, smartphone applications such as Firefox, Skype, Microsoft Office, and many others (Windows, UNIX, Linux, and many others). The challenges generated by the technologies are identical since they are comparable. The topic of cybersecurity has become well established in urban countries, and it is a significant issue. Many issues include critical information infrastructure (CII), SCADA services, government networks, and other issues linked to the internet infrastructure and user devices such as desktop computers, smartphones, and internet applications.

Cybersecurity challenges are treated as national security because from small to large organizations, public to private universities, and hospitals are faced with cyberattacks. From the Wikipedia analysis, most Middle East countries are affected by a massive attack called denial of service. Several cybersecurity challenges are presented, which are explained as follows.

- Advanced persistent threats
- Evaluation of ransomware
- Attacks through IoT devices or IoT threats
- Cloud security
- Attacks on blockchain technology and cryptocurrency
- Attacks designed via machine learning and AI technologies.

Advanced Persistent Threats

The advanced persistent threats are crucial cybersecurity challenges because they penetrate the system without providing any notification and stay in the system and server for a longer time, undetectable by anybody. These cyberthreats' main intention of stealing high-sensitive information is also designed to target the target system. Most organizations are failing to safeguard their files and information from advance persistent threats. Then the life cycle of the advanced persistent threats is illustrated in Fig. 1.1.

Evaluation of Ransomware

Ransomware is one of the advanced persistent threat life cycle attacks. During this attack, malware entered the system and started to encrypt the files slowly. Once the files are locked, this attack demands the bitcoin because it's hard to track. The hacker establishes the decrypted key; only the user makes the payment. Therefore, ransomware is the leading cybersecurity challenge to executives, IT, and data professionals. The sample ransomware attacking process is illustrated in Fig. 1.2.

Attacks through IoT Devices or IoT Threats

The next cybersecurity challenge is attacked through IoT devices or IoT threats. The IoT device is interrelated with digital and computing devices that transmit the data via the network. The IoT devices have a unique identifier that is used to identify the specific IoT device. During the information transmission, several attacks have happened, which will affect the securities. The insecure Wi-Fi, web interfaces, lack of security knowledge, and insufficient authentication methods are reasons for these security challenges.

Cloud Security

Cloud security is one of the major issues because it has been stored in a third-party server that requires high-security measures. The organizations create their cloud storage to store their data, but they fail to maintain their data security, leading to

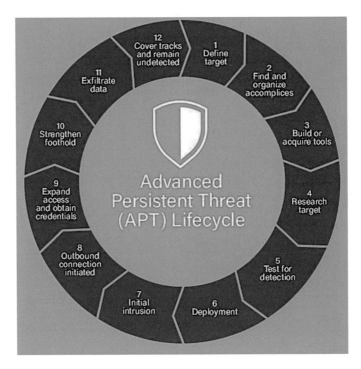

Fig. 1.1 Advanced persistent threat life cycle

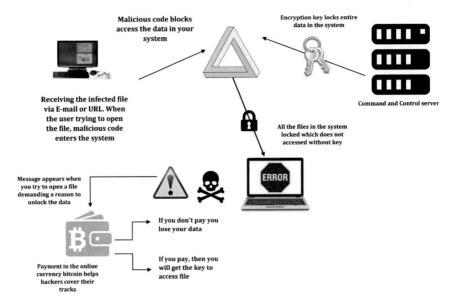

Fig. 1.2 Evaluation of ransomware

several cloud attacks. These cloud attacks are happened because of specter vulner-abilities, meltdown, insecure API, natural disaster, data loss, human error, and misconfigurations.

Attacks on Blockchain Technology and Cryptocurrency

The blockchain and cryptocurrency technologies are now at being level that requires adapting many of the organizations. Before undertaking these technologies, few security measures, authentication methodologies, and authorization procedures are needed to be updated. There are several attacks like DDoS attacks, Sybil attacks, and eclipse attacks that happen via these technologies.

Attacks Designed via Machine Learning and AI Technologies

Machine learning techniques and the artificial intelligence system use big data to decide on their application. Due to the growth of these technologies, hackers are quickly affecting the learning system using sophisticated attacks.

1.4 Cybersecurity Threats

Cybersecurity threats may typically be categorized into two general groups, including acts aimed at damaging or damaging network attacks and activities that attempt to exploit the cyberinfrastructure in illegal or harmful ways without jeopardizing or damaging cyber vulnerability infrastructure. Although certain intrusions do not immediately impact a cyber network's activity, for example, when a Trojan horse reaches the computer, such intrusions are known as cyberattacks. They can then enable acts to obliterate or corrupt the capability of the computer. Computer usage includes using the internet and other computer networks in theft, cheating, recruiting and educating terrorists, violations of copyright, and other laws limiting content distribution, distributing controversial communications.

1.5 Cyberattack Life Cycle

The exceptional growth in cyberspace has generated incomparable economic growth, wealth, and prosperity. It invites fraudulent persons with entirely new risks and crimes. Industry and policy priorities are fundamentally linked to maintaining and facilitating cyber-based transactions and activities. For financial growth, wealth, effi-ciency, governments need safe global digital infrastructure. The cyberattack life cycle includes research, developing, delivering, exploiting, infiltrate, C2, and act. Figure 1.3 shows the cyberattack life cycle.

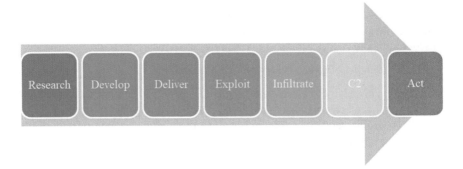

Fig. 1.3 Cyberattack life cycle

1.6 Cybersecurity Principles

The Information Technology Industry Council (ITI) establishes a comprehensive collection of cybersecurity standards for both business and government. Throughout ITI, the world's most significant development firms are information software and services suppliers and customers. To strengthen cybersecurity, ITI has established six principles. In addition to surviving, the organization must make efforts to maintain information security:

- Organizations must focus on public–private collaborations and develop their current capital investments and initiatives. The IT industry has been leading, providing resources, innovation, and stewardship in all aspects of cybersecurity through a partnership with the government over many years. When exploiting and expanding on such current initiatives, investments, and collaborations, information security strategies are highly successful.
- The high, connected, and multinational complexity of the network communication world is expressed in the organizations. Cyberspace is a unified and international system extending geographic boundaries and crossing national jurisdictions. Countries will practice leadership in encouraging the usage of guidelines, best practices, and services for the development of security and interoperability within the sector and globally accepted.
- Firms must evolve rapidly and focus on effective risk control to respond to new challenges, technology, and business models. Cybersecurity strategies will be focused on risk assessment. Security is a device for realizing and ensuring continued trust in various cyberinfrastructure techniques. Cybersecurity initiatives will assist a corporation in identifying, assessing, and implementing action to mitigate emerging threats.
- Cybersecurity efforts should focus on awareness. The information protection principle is to increase the understanding of the public. The owners of cyberspace involve customers, businesses, governments, and the owners and operators of

infrastructures. Such stakeholders need to be encouraged to track information security activities and recognize their important position in resolving such threats and risks to their properties, reputations, businesses, and companies.

- Ensuring information security has to focus more specifically on bad actors and their risks. Cyber attackers have now provided whole new opportunities for theft because of the centralized, interconnected, and interactive internet technology design. Security activities overcome these incentives and allow for cyber transactions and activity.
- In cyberspace, like in the real universe, adversaries utilize fraud, surveillance, or warfare tools. Cyber protection strategies will enable policymakers to leverage established laws, activities, and knowledge exchange processes to respond domestically and internationally to cyber entities, threats, and incidents.

1.7 Cybersecurity Standards

In a dynamic era, cybersecurity standards are different from each other because each region follows specific standards. The cybersecurity standards have some rules based on these organizations or institutions processing their data. For example, patient data, online payment processing are utilized with different regulations to complete their task. The specific standards are more useful in maintaining their data security. In addition to this, cybersecurity standards are defined with a set of policies that are applied by a particular company. These policies help prevent vulnerabilities while making any transactions like online payment, data transmission, etc. Therefore, most organizations follow their standards to improve their security level. Here few cybersecurity standards are explained.

ISO 27001

The ISO 27001 is a general cybersecurity standard utilized by almost every organization to manage its data security. This standard consists of several rules and procedures approved by the organization by conducting different tests and certified. Based on security standards, organizations need to continuously support and update the technologies and servers to eliminate the vulnerabilities. It is the international standard to use the ISMS policy to utilize the ISO 27001 standard.

PCI DSS

Payment Card Industry Data Security Standard (PCI DSS) is utilized by the organization to support the payment gateway. With the help of this, standard business process saves the card details and user information; also, the secured payment process is initiated. This standard is created by the different card brands, and organizations must update their technologies to predict security-related issues.

HIPAA

Hospitals support the Health Insurance Portability and Accountability Act (HIPAA) related to cybersecurity standards. The HIPAA standard is utilized to protect the patient's health information from unauthorized access. During this process, the hospital must have strong network security with HIPAA standards to manage its critical health information. With HIPAA's help, medical information is transmitted by performing the encryption process to protect the maximum-security data.

FINRA

Financial Industry Regulatory Authority (FINRA) standard supports the financial bodies. During the funds and financial transactions, the FINRA standard is supported by several organizations to protect the data from illegal access. This is the most critical standard because every organization should comply with this FINRA.

GDPR

The European Government created the general Data Protection Regulation (GDPR) standard. This standard developed to protect every user's data and manage data security, authentication, and authorization. Therefore, the GDPR standard focuses on user safety, and every organization should comply with this standard to protect their data. Almost every organization and business follow a set of rules and policies for improving data security from the discussion.

1.8 Cybersecurity Framework

As discussed above, cybersecurity standard organizations should follow policies and rules to accomplish their data protection. The created policies are needed to certify by audit report to finalize the framework. According to businesses or organizations, the requirement cybersecurity framework is designed to manage their business security. There are different standards, such as ISO 27001, HIPAA, GDPR, PCI DSS, and FINRA standards, to protect the data. Based on the purpose, the framework is generated, and respective operations are performed. During the cybersecurity framework creation, it has few components that are listed as follows,

- Core
- Implementation tiers
- Profiles.

It ties in with executing the association's safety efforts, so the business coherence should be kept. The association must adhere to the specific arrangement of decisions that falls under a particular structure to actualize it. A few things must be considered, like the foundation ought to be secure; there should be no weaknesses in the framework; the product is used to ensure the framework should be refreshed. Any association that provides it follows the whole arrangement of strategies character-ized under the particular system is viewed as acceptable to execute the cybersecurity structure.

1.9   Fundamentals of Cybersecurity

At this age of rapid communication technology, change, and improvement in the world, networks, and systems, especially organizations, are becoming more inter-connected. Through traditional networks, it builds more digital infrastructure, gener-ates large amounts of internet data, migration to cloud providers, and provides third parties with better access to the data. To add this, companies have several more computers, as in the previous years, where the network is not organized, fragmented, and distributed, and only the office-based computer is used. This introduces organi-zations to more significant risks and vulnerabilities from cyberattacks, threats, theft of data, and so on. Throughout this decade, numerous violations and cyberattacks have led large organizations to lose their money, integrity, and brand identity. It has indicated that individuals and companies require complex information protection strategies. The attacks grow every day as attackers become more innovative and essential for properly defining cybersecurity and recognizing cybersecurity basics.

1.10   Why is Cybersecurity Important?

The reasons why cybersecurity in the dominant digital environment is so essential are listed below:

- The number of threats is rising rapidly every year. According to the McAfee study, cybercrime is now more than $400 billion, relative to $250 billion two years ago.
- Cyberattacks can be challenging for businesses to sustain. A data violation may cause immense reputational harm as well as financial loss to the company.
- Cyberattacks are increasingly destructive these days. More specialized forms are used by cyber attackers to initiate cyberattacks.
- Regulations such as GDPR force companies to excellent support for their data.

Due to the above reasons, cybersecurity has become a critical part of the business. The emphasis is now on developing adequate recovery systems that minimize harm if a cyberattack occurs. However, an organization or person who has a firm grasp of cybersecurity principles will establish a proper response strategy. Now that it knows what cybersecurity is and why it is essential to look at primary cybersecurity objectives. Figure 1.4 shows the fundamentals of cybersecurity.

Confidentiality, integrity, and availability are a paradigm developed to help organi-zations and entities to establish their security strategies, often known as the CIA triad. Cybersecurity means technically securing information from unauthorized access,

Fig. 1.4 Fundamentals of
cybersecurity

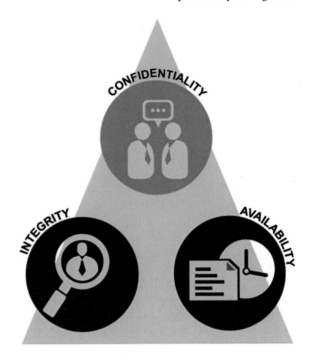

unauthorized change, and unauthorized deletion to maintain confidentiality, integrity, and availability. Let us discuss these components and other information protection steps to ensure the safety of each element.

1.10.1 Confidentiality

Confidentiality ensures that unauthorized parties are protected from the disclosure of data. It means trying to protect privately and securely the identities of the authorized parties involved in transmitting and retaining data. Cracking inappropriately encrypted files, man-in-the-middle (MITM) attacks, and the leakage of sensitive data often compromise confidentiality.

The standard confidentiality measures include:

- Encryption of data
- Authentication of two factors
- Biometric tests
- Tokens for security.

1.10.2 Integrity

Integrity requires the security of data from unauthorized parties to be updated. Information and services are only needed to be modified in a specified and authorized manner. Challenges that could threaten Integrity include turning a computer into a "zombie device" or introducing malware into web pages.

The standard integrity measures include:

- The standard integrity measures include:
- Value of cryptography
- Using file allowances
- Supplies of uninterrupted power
- Backups of data.

1.10.3 Availability

The availability ensures that approved parties may access the details when needed. Knowledge has value because it is available to the right person at the right time. Occurrences, including DDoS assaults, equipment malfunction, system failures, and human errors, may lead to data unavailability. The standard availability measures include:

- Data protection on hard drives
- Firewalls are implemented
- Supplies of backup power
- Quality of the data.

All cyberattacks could threaten one or more of the three CIA triad categories. Confidentiality, integrity, and availability will all work together to protect the information. It is essential to learn what the CIA triad is, how a strategy on quality security is structured and executed while knowing its different concepts.

1.11 Why Do We Implement Cybersecurity?

Numerous methods are available to implement cybersecurity, but three main steps address a security problem. Figure 1.5 shows the steps to treat cyberattacks.

The first step is to identify, for example, the problem of security when a denial-of-service attack or a man-in-the-middle attack occurs. The next step is to protect and detect the issue. All data and information which could be affected by the attack must be separated. Once the problem has been detected and responds, the last move is to recover a patch that successfully fixes the issue and restores the organization to a working state. Figure 1.6 shows the vulnerability, threat, and risk concepts.

Fig. 1.5 Steps to treat cyberattacks

Fig. 1.6 Vulnerability, threat, and risk

Three criteria are held in mind for various assessments: identification, analysis, and treatment of a cyberattack. It is mentioned as follows:

- Vulnerability
- Threat
- Risk.

1.12 Cybersecurity Attacks and Their Types

A cyberattack is a type of aggressive action that seeks to hack, modify, or kill data or information systems utilizing database programs, infrastructures, computer networks, or computers in different ways. A cyberattack is a movement of any kind employed by persons or organizations in their entirety infrastructure, electronic

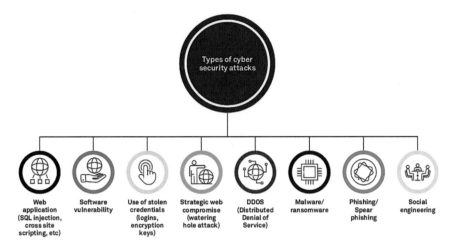

Fig. 1.7 Types of cybersecurity attacks

information systems, and networks of computers and personal computers networks computers and personal computers from a source that robs alter, or destroys anonymous a target specified by a susceptible system hacking. Cyberattacks may involve the installation of spyware on a computer. They are trying to destroy the whole nation's infrastructure. An attacker is an individual or process trying to access data functions or other device-limited areas without permission with malicious intent potentially. Cyberattacks may form part of cyber warfare or cyber terror, depending on the context. Figure 1.7 shows the types of cybersecurity attacks.

1.12.1 Malware Attack

The attack by malware is a typical cyberattack in which malware conducts unauthorized actions on the system of the user. The malicious malware (named the virus) includes a range of different attack types, such as ransomware, spyware, control, and more. The use of malware has been accused (and sometimes caught) of criminal organizations, state actors, and best-known businesses. As in most cyber threats, several ransomware hacks have world media coverage due to their significant impact. Although there is a discussion about whether or not ransomware victims can pay off, some businesses have become threatening enough to purchase bitcoin preventative measures even if they get hit with ransomware and decide to buy it. Figure 1.8 shows the malware attacks.

Types:

- Trojan horse
- Virus
- Worm.

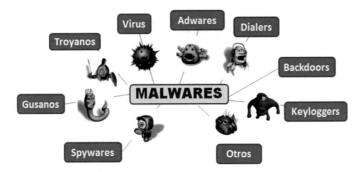

Fig. 1.8 Malware attacks

1.12.2   Phishing Attacks

Phishing is a cyberattack by software development that aims to distribute sensitive/precious information to targets. Sometimes referred to as phishing fraud, attackers target the person's login information, financial records (such as credit cards or bank accounts), business details, etc. Because of their sheer size and their ability to find holes in their security systems, large organizations risk long-term phishing threats. An employee who targets the phishing attack will attempt his company if this attack is completed. Organizations will determine phishing attacks' vulnerability through systematic monitoring and implementing training programs for security knowledge.

Types:

According to its fundamental definition, the term phishing attack sometimes ensures a broad attack on many users (or targets). The attacker should take very little preparation to do that in the expectation of at least some of the objectives being affected, making the minimum flow effort attractive. However, the expected benefit for an attacker usually is not all that great. It can be seen as a "quantity over quality" approach. Figure 1.9 shows the phishing attack.

- Spear phishing
- Whaling
- Clone phishing.

1.12.3   Denial-of-Service Attacks

A denial-of-service (DoS) attack is targeted at blocking a network or utility by inserting artificial traffic through a target that limits consumer access to the device to be attacked. Denial-of-service (DoS) attacks concentrate on preventing or blocking

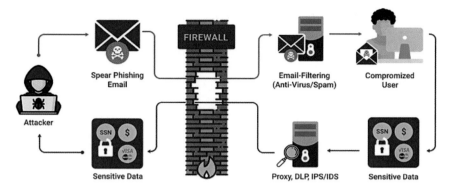

Fig. 1.9 Phishing attack

the connection of authorized users to websites, applications, or other services. Criminal organizations have used those attacks, activists to extort money, and state actors to threaten their adversaries. Figure 1.10 shows the denial-of-service (DoS) attack.

Types:

- Distributed denial of service (DDoS)
- Network-targeted denial of service
- System-targeted denial of service
- Application-targeted denial of service.

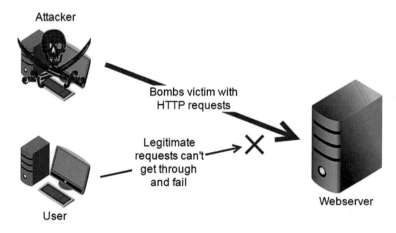

Fig. 1.10   Denial-of-service (DoS) attack

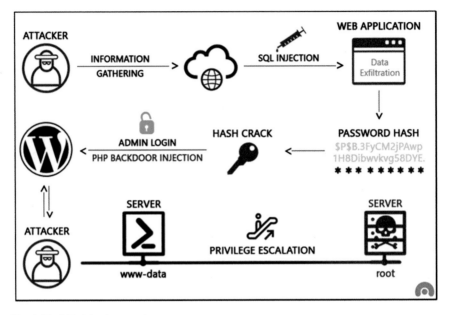

Fig. 1.11 SQL injection attacks

1.12.4 SQL Injection Attacks

Structured Query Language (SQL) is a language built in a database to process and analyzes data. SQL has gradually entered various industrial and open source databases since its creation. SQL injection (SQLi) is a form of cybersecurity attack that uses a specific SQL statement to trick systems into unnecessary and unexpected things. Figure 1.11 shows the SQL injection attacks.

Types:

- Unsensitized input
- Blind SQL injection
- Out-of-Band injection.

1.12.5 Cross-Site Scripting

Cross-site scripting is an attack on a vulnerability in the web-based injection system, which delivers harmful scripts to the web server. Targets are not attacked by each other, but somewhat vulnerable websites and web applications are used as users connect with cross-site attacks with such sites/applications. Cross-site scripting is an attack on a vulnerability in the web-based injection system, which delivers harmful scripts to the web server. Targets are not attacked by each other, but somewhat vulnerable websites and web applications are used as users connect with cross-site attacks with such sites/applications.

For instance, an unsuspected user will visit a compromised web page, and the attacker loads and performs a malicious script through the browser. This will take inspiration from infiltration/theft, hijacking, and more. JavaScript is ubiquitous in XSS attack developers since several web browsers and platforms are commonly available, although any browser-based language can be used to develop the attack. While they have been around for more than 15 years, XSS attacks are highly effective and frequently used as a growing and viable attack vector these days.

Types:

- Reflected XSS
- Persistent XSS
- Dom-based XSS.

1.12.6   Man-In-The-Middle Attacks

Man-in-the-middle (MITM) attacks are a standard method that allows attackers to avoid communication between two targets. The assault happens between two hosts who connect legally and will enable the attacker to "listen" to a discussion that they usually cannot understand, thus the name "man in the middle." Figure 1.12 shows the man-in-the-middle attacks.

Types:

- Rogue access point
- ARP spoofing
- mDNS spoofing
- DNS spoofing.

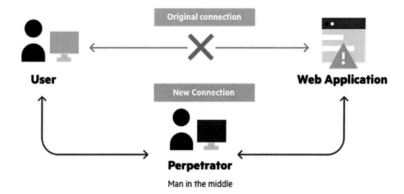

Fig. 1.12 Man-in-the-middle attacks

1.13   Cybersecurity Development

Cyberspace is a universal and interrelated area that covers national and geographical boundaries. IT companies continuously innovate and contribute to producing
different products and services to foster growth, activity, maintenance, and safety
in this field. The cyberspace users, including customers, corporations, governments,
technology owners, and operators, search for a reliable, stable cyberspace experience. Cybersecurity efforts should reflect cyberspace's borderless nature and should
be based on internationally established standards, best practices, and international
programs for assurance. Cybersecurity standards enhance digital infrastructure interoperability since security systems and technologies can better be connected across
borders.

1.14   Advantages of Cyber Security

Cybersecurity has several advantages because it provides security to systems and
networks while making data transactions.

- Organization security

  - Cybersecurity protects the organization's systems and network from external
    attacks. So, that organization feels safe from unauthorized access and cyber-
    attacks.

- Safeguard for sensitive data

  - Cybersecurity manages sensitive data like patient health details, sales details,
    student details, and bank transactions from unauthorized access.

- Unauthorized access hamper

  - The cybersecurity process restricts unauthorized users from accessing the data.
    The shared or stored data is only accepted when the user is authenticated using
    respective authentication details like OTP, password, fingerprint, etc.

- Data Reliability

  - Once the organization utilizes the cybersecurity concept, its data has become
    more reliable, and the data is used for the further analysis process.

1.15 Applications of Cybersecurity

Here few cybersecurity applications are included:

- Cybersecurity is widely utilized in business applications because it protects data from phishing, malware, social engineering, and ransomware.
- It protects the network and data
- Cybersecurity is utilized in digital assets because it helps to terminate unauthorized access.
- The security concept is used to protect end-users sensitive and personal information.
- Increases organization confidence while processing the data.

Summary

This section explained that the introduction of cybersecurity, fundamentals, challenges, and the importance of cybersecurity will be given the beginner's knowledge and explained how cybersecurity can be implemented and how the cyberattack can be controlled. The next section will provide the basics of hacking and penetration test.

Preview

This chapter covers the basics of hacking and their introduction to hackers' classification. It gives a detailed overview of ethical hacking and types of cybercrime, DoS. Finally, it provides a deep explanation of the penetration test and penetration test phases with a security assessment.

Assignment Questions

1. Write down the security challenges in cybersecurity?
2. Explain in detail cybersecurity standards and framework?
3. Why is cybersecurity important, and how is it implemented?
4. Write down a detailed explanation of cybersecurity attacks?
5. List out the application of cybersecurity and case study about any one application.

Multiple Choice Questions

1. Trojan horses are similar to a virus in the matter that they are computer programs that replicate copies of themselves

   (A) True (B) False
2. ……… monitors user activities on the internet and transmits that information in background to someone else.

   (A) Malware, (B) Adware (C) Spyware (D) None of these
3. Unsolicited commercial email is known as ……

   (A) Spam (B) Malware (C) Virus (D) Spyware

4. Which of the following is a class of computer threat?

   (A) Phishing (B) DoS attack (C) Soliciting (D) Stalking
5. Which of the following is not an external threat to a computing or computer network?

   (A) Ignorance (B) Adware (C) Trojan Horses (D) Crackers

Answers

1. False
2. Spyware
3. Spam
4. DoS attack
5. Ignorance.

Summary Questions

1. What is the plan for identifying and addressing cyber threats?
2. What is the top cyber security concern?
3. What is the CIA?
4. What is a Firewall?
5. Explain the brute force attack. How to prevent it?

Chapter 2
The Basics of Hacking and Penetration Testing

2.1 Hacking Introduction

Hacking is a crack inside a computer to use a private network or computer system. In short, for some illegal purpose, it is unlawful to control or access computer network security systems. Computer hacking mentions the preparation of altering or transforming computer software and hardware to achieve an objective measured beyond the creator's original purpose. Those people engaged in computer hacking are typically known as "hackers." Most hackers have advanced computer technology knowledge. The typical computer hacker is skilled in a particular computer program and has advanced computer programming capabilities. Computer hacking is somewhat tricky and ambiguous to define, unlike most computer crimes, considered a clear cut in legality. However, in all forms, computer hacking entails a certain degree of breach of others' privacy or damage to computer-based possessions like software, web pages, or files. Figure 2.1 shows cybersecurity attacks and hacking.

Hacking can separate into various groups depends on what is being hacked: website hacking, ethical hacking, network hacking, password hacking, computer hacking, and email hacking. The benefit of hacking is moderately valuable in the following situations: to execute penetration testing to support network and computer security, put suitable preventative actions to avoid security breaches, and recuperate lost data, particularly in case of a lost password. Hacking is unsafe if it is prepared with harmful commitment. It can impact denial-of-service attacks, massive security breaches, hampering system operation, privacy violations, unauthorized system access to private information, malicious attacks on the system.

Hackers are classified following their exploits' intentions. The incline below categorizes hackers as projected.

- **Ethical Hacker (White hat)**: A hacker that has access to systems to remedy the deficiencies. Penetration testing and vulnerability assessments can be carried out, as well.
- **Cracker (Black Hat)**: A hacker who gets unapproved personal gain access to computers. Usually, the purpose is to violate data, snip corporate data, protection

© The Author(s), under exclusive license to Springer Nature Singapore Pte Ltd. 2023   21
B. S. Rawal et al., *Cybersecurity and Identity Access Management*,
https://doi.org/10.1007/978-981-19-2658-7_2

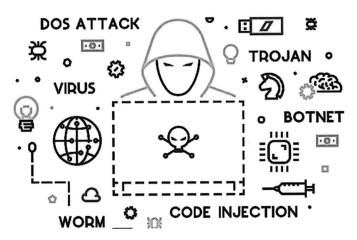

Fig. 2.1 Cybersecurity attacks and hacking

rights, and transfer amounts from bank accounts. Black hat has always been illegal because of its lousy intention to rob, violate privacy, corporate data, damage the system, block communication to the network, etc.

- **Gray Hat**: A hacker between black and ethical hackers. To predict disclosure and weaknesses them to the system owner, its disruptions in computer systems without authority.
- **Script Kiddies**: A non-skilled person with already built-in tools who can access computer systems.
- **Hacktivist**: A hacker who utilizes hacking to send messages from society, religion, politics, etc. This usually occurs through the removal of websites, and the message is left on the hijacked site.
- **Phreaker**: A hacker that recognizes and activities the phone and computer weaknesses.

2.2 Ethical Hacking

Ethical hacking detects computer system vulnerability and computer network vulnerabilities and offers countermeasures that defend vulnerabilities. The following guidelines will apply to ethical hackers: seek written authorization before hacking from the computer device vendor and computer network. The organization's privacy is secured. All found vulnerabilities of the computer system are disclosed transparently to the client details on discovered vulnerabilities to hardware and software vendors. Data is one of the organization's most significant assets. Maintaining protected information will protect the identity of an entity and save money for an organization. Hacking could lead to business losses for financial organizations like PayPal. Ethical hacking puts cybercriminals a step ahead, which would otherwise result in business loss. Moral activity is lawful if the hacker complies with the rules.

Cybersecurity is an information technology field that relates to networks and devices and is an area that is part of the broader computer security context. The goal of this field is to limit computer crimes, especially hacking and identity theft crimes. The cybersecurity region is attempting to shield information and corruption to avoid the danger of computer crimes. Moreover, the personal blocking of explicit or pornographic content accessible on the internet is a unique protection method. The blocking of this content is carried out personally through the internet portal of the individual. Cybersecurity typically refers to collective structures and processes that secure sensitive information and resources against publication, tampering, or a range of unwanted activities planned and implemented by unsuspecting individuals or unplanned activities. Cybersecurity programs incorporate various techniques to prevent inappropriate computer behavior so that unauthorized access can be reduced.

The primary use of computer safety is the technology used to secure an operating system. This technology was founded in the 1980s and used to manufacture impenetrable operating systems. While it still uses this technology today, system management changes have been imposed, which have resulted in greater complexity. These versions are based on the operating system's kernel technology that ensures that some security policies are assessed in a particular working environment. This plane is based on a combination of distinct microprocessor functions that often involve memory management in a correctly applied kernel. This is essentially the basis for a secure operating system that ensures that hostile elements do not penetrate.

Cybercrime consists of networks and computers for illegal activities like computer virus transmission, unauthorized transferable electronic funds, online bullying, etc. The majority of cybercrimes happen on the internet. That cybercrime can be committed via SMS and online chat applications using mobile phones.

Ethical Hacking Process

In every IT project, ethical hacking must be planned earlier to manage the security factors. During this process, tactical and strategic issues are determined while performing ethical hacking. Therefore, planning is more critical because a simple password-cracking test includes several penetration tests in web applications.

Plan Formulation

Formulating a plan is an essential task in ethical hacking approval. This process should be visible and known to everyone in the project; it must be visible to decision-makers. For every proper hacking action, the backup should be maintained in your manager, customer, executive, and boss because it helps authorize your tests. The authorization is done by making a simple memo from your management that requires a while performing test in your system. If the user wants to complete the testing process, the contract approval helps to authorize the customers. The formulating plan includes the following scopes.

- Particular systems need to be tested

- The risk involved while testing the system
- What tests are performed over the timeline?
- How the tests are accomplished?
- Listed the significant discovered vulnerabilities
- The higher-level report outlining vulnerabilities to be identified during the implementation.

Tool Selection

Tool selection played an essential role in ethical hacking because if you select the wrong tool, the real moral hacking accomplishment is so tricky. The chosen tools must cover the technical and personal limitations while discovering the vulnerabilities. For example, if you are doing the physical security assessment, many tools miss the vulnerabilities that only focus on the specific test instead of analyzing everything. Therefore, the selected tools must focus on a task, questioning, and everything in the application that improves the ethical hacking process. The chosen tools must have the following features.

- Acceptable documentation
- Need to provide a detailed report regarding the discovered vulnerabilities with how they are fixed and exploited
- Tools must be updated and support the current version
- High-level reports are submitted to the managers or non-technique types.

Plan Execution

Time and patience are more critical in ethical hacking while executing the tests. The hackers are continuously monitoring the user activities to accessing the information from your system. Therefore, vital and private information should be managed as possible. Primarily, the generated test reports are needed to be maintained securely from the hacker. The security has been established via email encryption that is done by using pretty good privacy.

Result Evaluation

After executing the test, the results are needed to assess the discovered vulnerabilities and uncovered vulnerabilities while performing the test—the result evaluation process correlating the discovered vulnerabilities with user experience. The examined results are forward to the upper management for evaluating the process.

2.3   Types of Cybercrime

- **Computer fraud**: Intentional deception in the use of computer systems for personal benefit.
- **Identity theft**: Stealing and impersonating personal details from others.

- **Copyright files/info sharing** includes exchanging protected copyright files like eBooks, computer programs, etc.
- **Transfers of electronic funds**: This includes unauthorized access and illegal money transfers to bank computer networks.
- **Electronic money laundering**: This contains the usage of a computer to launder money.
- **ATM fraud**: Steal information of an ATM card like account and PIN. These specifics are utilized to withdraw funds from the accounts seized.
- **Denial-of-service attacks**: This requires using multilocated computers to attack servers to shut them down.
- **Spam**: Sending unlicensed emails. Typically, such emails consist of advertising.
- **Hacking**: Intruder accessing the system without having user permission and access the data.
- **Virus dissemination**: Viruses are attached to the file to damage the system and circulate the other computer or network.
- **Logic bombs**: It is called slag code or malicious code inserted into the software and executes the malicious task triggered by a particular job.
- **Phishing**: It is done by email spoofing and retrieving the user's confidential information.
- **Email bombing**: A bunch of emails is continuously sent to the user and crashes the user accounts.

Any secret method of bypassing regular authorization or security controls is a backdoor in the computer system, cryptosystems, or algorithms. They may exist for a number of reasons, whether original or poorly designed. They could have been added to permit specific legitimate access by an authorized party or by an attacker for malicious details; they create a regardless vulnerability of the reasons for their existence. Backdoors may be challenging to detect, and backdoor detection is usually detected by someone who has access to the app source code or confidential information of the OS of the computer.

Cryptography is a method for protecting information and communication using codes to be read and processed only by people for whom the data is intended. Cryptography refers to computer sciences to protected data and communication technology derived from computational principles and a series of regulatory equations called algorithms. Messages can be transformed in ways that are difficult to decode. These powerful algorithms are utilized to generate cryptographic keys, digital signatures, data security authentication, online browsings, and hidden communications, like email and credit card transactions. The following four objectives concern modern cryptography, confidentiality, no one for whom the information was unintended can understand, integrity: information cannot be altered between the sender and destination recipient in storage or transit without the identification of the alteration, non-repudiation at a later point, the creator/sender of the information can not dispute the existence or transmission of the data, authentication: the sender and the recipient should check the identity of each other and its origin and destination. Cryptography is categorized into two types, asymmetric key encryption or public key encryption and single key or symmetric encryption.

Figure 2.2 shows asymmetric cryptography or public key cryptography. Asymmetric cryptography is an encryption system using keys pairs: public keys that can be broadly distributed and private keys that can only be known to the user. These keys are produced based on cryptographic algorithms, which have single-way functions based on mathematical problems. Appropriate security only requires the private key; without compromising safety, public access can be openly circulated. Any individual in such a system can encrypt the message using the public key of a recipient, and the private key of the receiver can decrypt this encrypted message. This is essential even to secure authentication. The sender will use a personal resolution to merge a statement and create a small digital signature on the message. Whoever has the correct public key of the sender will combine the same message with the so-called digital signature to verify that the signature, i.e., made by the recipient private key owner, was valid. Public key algorithms are essential security ingredients in the secrecy, reliability, and unpredictableness of electronic communications and data storage in modern cryptosystems, applications, and protocols.

Figure 2.3 demonstrates symmetric key cryptography. Symmetrical encryption systems rely on a single key shared by two or more users. The same access is utilized to decrypt and encrypt the so-called plaintext (the message or data element being encrypted). The encoding method consists of running a plaintext (input) using an encoding algorithm known as the cipher and generating a ciphertext (output). The only way to read or view the code text's information is by using the correct key to decode it when the encryption scheme is powerful enough. The decryption process essentially transforms the ciphertext into plaintext. The protection of symmetric encryption systems is based on the corresponding key's random existence to force

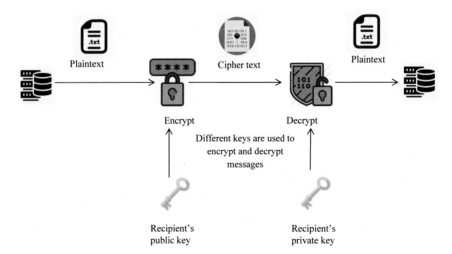

Fig. 2.2 Asymmetric cryptography or public key cryptography

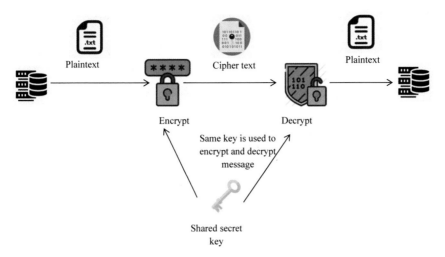

Fig. 2.3 Symmetric key cryptography

them through brute force. E.g., a 128-bit key will take billions of years to invent with popular computer hardware. The longer it is, the harder it is to crack. The more difficult it is. Keys 256-bit in length are commonly known to be highly stable and resistant to brute quantum machine attacks.

"Malware" denotes different types of harmful software, including ransomware and viruses. Once the malware has been placed on your computer, all kinds of havoc can occur, including taking control of the machine, keystrokes, and monitoring activities, silently forwarding all sorts of confidential information from the computer or network to the home base of the attacker. Attackers will utilize various approaches to get malware into their computers. At some point, however, the user often needs to take steps to install the malware. This may be by opening an attachment or snapping a link to download a file that might look innocent (like a PDF-Annex or a Word document).

2.4   Denial-of-Service Attacks (DoS)

Denial-of-service attacks (DoS) are intended to create the network or machine resource inaccessible to the intended operators. Attackers can deny service to specific victims, like deliberately entering an incorrect password for the victims to be locked, overloading a machine or networkability, and simultaneously blocking all users. The attacker can stop the service. While a network assault from a single IP address is congested by adding a new firewall instruction, several forms of DDoS attacks, where the attack comes from a high number of points, are possible—it's much more challenging to defend. These attacks may arise from botnet's zombie computers or

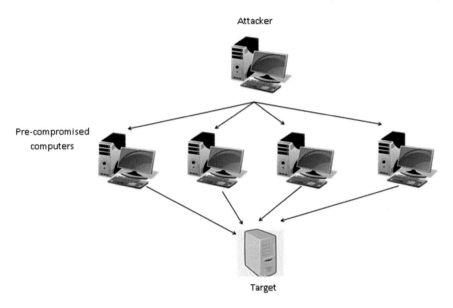

Fig. 2.4   Denial of service (DoS)

other possible techniques, such as reflections and amplification, which mislead inno-cent systems into sending the victim's traffic. Figure 2.4 shows the denial-of-service attacks (DoS).

Distributed denial-of-service (DDoS) is a large-scale DoS attack those custom more than one internet protocol address or computer from malware-infected hosts. Typically, a DDoS attack includes about 1–5 nodes on different networks; fewer nodes may be called DoS attacks and are not DDoS attacks. Since the incoming traffic floods from various sources originated, the attack could not be stopped simply by filtering in. It makes it difficult to differentiate between authentic user traffic and aggression when spread across multiple sources. Application layer attacks use DoS-related activities and can impact server-running software to use all CPU time or memory or seal the available disk space. Attacks can saturate endless resources with particular packet types or link requests. An attacker's primary advantage is that multiple machines can produce high attack traffic than a machine. Numerous attackers can be tougher to turn off than one assault machine, and that every attack machine's behavior is harder to track and shut down. This attacker benefits impact defense mechanism challenges. For instance, it may not help just buy more incoming bandwidth than the attacker, as the attacker may only add more attackers. After all, this will crash a website for several periods. The emblems of DDoS attacks involve the website being unresponsive. The website is reacting slowly, and the user has issues accessing the website and internet connection issues if you are a target. A DDoS attack would deplete the server and increase the load time for the website. If a DDoS attack hits a website, performance problems can happen, or the server can crash, irresistible computer resources like memory, CPU, or even the whole network. Most attacks by DDoS nowadays come

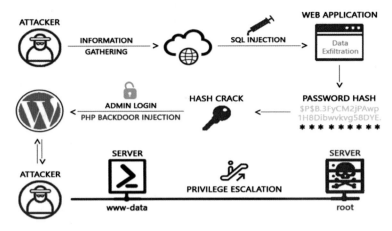

Fig. 2.5 SQL injection attack

from a botnet of IoT vulnerable devices operated by hackers. It includes internet-based surveillance cameras, kitchen apps, smart televisions, home lighting systems, and even fridges. DDoS attacks' exponential growth has been mainly due to an overall lack of control over the Internet of Things devices, which makes them outstanding botnet workforces. Hijacked clusters of the Internet of Things devices can redirect malicious applications to sites that cause the attack of DDoS with single IP addresses. To combat phishing attempts, we must understand how important it is to verify email and attachments/links.

Figure 2.5 shows the SQL injection attack. A threat based on push uses spam, phishing, or other wild means to attract a user to a malicious website (often rummaged) and collects information, and injects malware. Phishing, DNS poisoning, and other methods appearing to come from a trusted source are the basis for push attacks. An attack from the SQL injection system uses any known vulgar SQL, enabling malicious code execution by the SQL server. If, for instance, a SQL server is defenseless to a firewall attack, an attacker can go to the search and type in website code, forcing the SQL server to dump all of its password's usernames been saved for the site. Attackers can utilize vulnerabilities in SQL injection to circumvent security measures in applications. They can authenticate or authorize a web application or web page and recover the entire SQL database material. Any web application or website which utilizes SQL databases like Oracle, SQL Server, MySQL, or other users may be affected by SQL injection vulnerability. Criminals can use them to access sensitive information without permission: customers, personal information, trade secrets, IP, etc. SQL injection attacks are a web application vulnerability that is one of the oldest, most common, and most dangerous. For target, a SQL injection, a web application, or a web page must first identify insecure user inputs. Such user feedback is utilized openly in a SQL query for a web application or web page with SQL injection vulnerability. The attacker can generate input content. Such content, which is a significant part of the attack, is called a malicious payload. Only malicious SQL commands are executed inside the database after the attacker sends this information.

Figure 2.6 demonstrates the cross-site scripting (XSS) attack. One of the most common techniques to arrange a cross-site scripting attack is through malicious code injecting into a script or comment that can be executed. For example, a relation to JavaScript can be embedded in a statement on a blog. XSS could crucially damage a website's reputation by compromising users' data without giving any sign of maliciousness. Any sensitive data that a user sends to the website, like credentials, credit card data, and other personal information, can be hidden via XSS without the owners of the website comprehending that there was a single issue. In XSS, an intruder directly accesses the website without permission by hacking it anonymously and stealing cookies and clipboard contents remotely. Others will avoid this malicious behavior. The following recommendations help you prevent XSS attacks from users' input sanitation—a place that does not have the correct sanitization of the data in such a search area. This has to sanitize user data to capture potentially malicious user information. XSS HTML filter—an XSS Java filter—is used to sanitize user input for maliciously injecting HTML code properly. XSS protects a library that allows users to prevent attacks by cross-site scripting, allowing developers to eliminate all the potential for XSS attacks. HTML purifier, an HTML PHP filtering library, is used to delete malicious code from the input and plugin for most PHP frames. To identify XSS vulnerabilities, use web vulnerability scan tools such as Scan My Server, Site Guarding, Detective, SUCURI, and so on. Use web vulnerability scanning tools to identify web vulnerabilities.

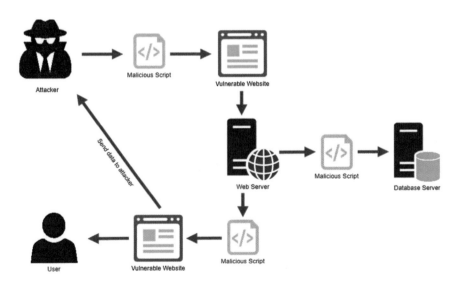

Fig. 2.6 Cross-site scripting (XSS) attack

2.5 Penetration Testing

A penetration test is a simulated, authorized cyberattack against computer systems known colloquially as a pen test or ethical hacking. The test identifies both weaknesses and the possibility of unauthorized parties having access to the system characteristics and information and strengths to complete the risk assessment in its entirety. This process typically classifies the target systems and a specific objective, examines the information available, and uses various methods to achieve this. A penetration test target can be a blank box (with system and background data) or a black box (with fundamental or no data other than the company). A gray box penetration test is a mixture of the two (where the auditor shares the target). An input assessment can help identify if a system is susceptible to attacks, if resistances are adequate, and which resistances (if any) the test has been overwhelmed. The system owner might be notified of security problems that the penetration test uncovers. The penetration test reports can evaluate potential organizational impacts and suggest risk-reduction countermeasures. A penetration test's purposes differ regarding what type of action is approved for any undertaking, with the primary objective of identifying vulnerabilities that a negative actor can use and informing customers of such openness and recommended mitigation strategies. This can help to recover from disasters and planning business continuity. It simulates intruders' methods of unauthorized access to the networked systems of an organization, compromises them, and uses open-source tools and proprietary to conduct the testing. Besides penetration testing, automated techniques require manual techniques for targeted testing of specific systems to guarantee that safety flaws have not been previously detected. Resources restrict the tester in the context of penetration testing: skilled resource, time, and access to equipment, as defined in the penetration testing agreement. Dynamic analysis of system configurations, network architecture, design weaknesses, technical defects, and vulnerabilities involves penetration testing. A penetration test not only indicates vulnerabilities and shows how weaknesses can be demoralized. On completion of penetration testing, pen testers provide executives, management, and technical audiences with an extensive report detailing the identified vulnerabilities and the suite of recommended countermeasures. An attacker is different only by intent, absence of malice, and authorization from an attacker.

Incomplete and unprofessional testing for penetration can lead to service loss and business continuity disruption. Therefore, without proper permission, employees or outside experts shall not conduct pen tests. Client organization management should give explicit authorization in writing for the conduct of penetration tests. This endorsement should contain a clear scope, a description of what to test, and when to do the test. Due to the nature of the pen tests, this approval's failure could, despite best intentions, lead to computer crime. Penetration tests are part of a comprehensive security audit.

In five phases, the penetration tests process can be simplified:

1. Scanning—Use technical tools to enhance the system information of the attacker. Nmap can be utilized, for example, for scanning open ports.

2. Reconnaissance—The act of collecting essential data on a target system. This can be utilized to attack the target better. Open-source search engines, for example, in search of information that can be used in the attack.
3. Access gain—The attacker can utilize a payload to activity the targeted system through the information gathered during the recognition and scanning phases. Metasploit can, for example, be used to automating attacks against recognized vulnerabilities.
4. Maintaining access—It needs to collect as much data as possible within the target environment.
5. Covering tracks—The attacker should clear any signs, any data collected, or log events of the victim's system to remain anonymous.

Figure 2.7 shows the penetration test methods. External penetration tests target an internet visible company's asset, for example, the web application themselves, the company's website, and the domain name servers. The goal is to gain access to valuable information and extract them. An internal test simulates an attack by a malicious insider in a test tester with access behind its firewall to an application. This isn't necessarily a rogue worker simulation. An operator whose credentials have been pinched because of a phishing attack may be a collective starting situation. In a blind test, the target company name is given to a test tester. This allows security

Fig. 2.7 Penetration test methods

operators to examine in real time how a real application assault would arise. Double-blind testing security personnel are not previously aware of a simulated attack in a double-blind test. In real time, formerly a tried breach, they will not have time to protect themselves. In this targeted testing, the samples maintain the security, cooperate and assess their movements. This is a respected training implementation that offers hacker feedback in real time to a security group.

Legal operations allowing the tester to perform an illegal action comprise SQL commands that are uninformed, unchanged passwords for visible source projects, human contacts, and old hazing or cryptographic features. A single flaw may not suffice to make a critical exploit possible. It is usually necessary to use many known defects and to form the payload as a valid operation. Metasploit offers a ruby library for everyday tasks and preserves a known exploit database. Fuzzing is a common technique for discovering vulnerabilities under budget and time constraints. It seeks a random input to obtain an untreated error. To access less often-used code paths, the testing system uses random information. Well-traveled code paths are generally bug-free. Errors are beneficial because they either disclose additional data, like full trace backtrack HTTP server crashes, or are directly usable, such as buffer overflows.

The following activities guarantee an excellent penetration test, creating the penetration test variable such as goals, limitations, and procedural justifications. Refer to qualified and skilled professionals for the test appointment of a legal penetration tester, which follows the disclosure agreement rules. Choose an appropriate set of tests to balance costs and benefits. For the following reasons, penetration testing is essential for companies identification of the threats to information assets of an organization. Reduce an organization's IT security expenditure and improve safety investment returns by identifying and mediating vulnerabilities/weaknesses and ensuring comprehensive security assessment involving procedure, policy, implementation, and design. Certification and maintenance of industry regulation. The adoption of best practices in compliance with legal laws and industry. Efficacy testing and validation of safeguards and controls. Modification or improvement of existing software, hardware, or design infrastructure. They were targeting serious vulnerabilities and stressing security issues at the application level for management and development teams. Provide a comprehensive method to preparatory steps to prevent future exploitation. Assess the effectiveness of network protective equipment, e.g., firewalls and routers and servers for the web.

A security audit will only verify that the organization follows a number of standard safety policies and procedures. It is a systematic method of technical appraisal of an organization's system that comprises manual interviews with employees, conducting security scans, security checks, and physical analysis of access to organizational resources in various access controls. A vulnerability analysis focuses on discovering information system vulnerabilities. It does not show whether the vulnerabilities can be exploited or how much damage could be the successful vulnerability exploitation outcome. Penetration testing is a security assessment methodological method that includes an audit for security and evaluation of vulnerability and shows whether attackers can successfully use the system's vulnerabilities.

Fig. 2.8 Vulnerability
assessment

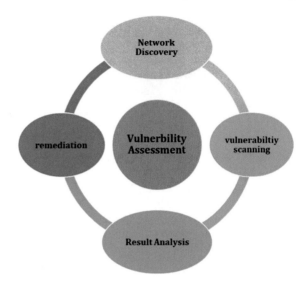

2.5.1 Penetration Testing Versus Vulnerability Assessment

Penetration testing and vulnerability assessment are different from each other. Penetration testing is the action to perform while happening during the internal and external cyberattacks. During this process, the ethical hackers or penetration testers utilize different tools to analyze the sensor data and control the critical activities. Invulnerability assessment, security vulnerabilities are detected in the environment. In this process, the result has been analyzed to perform the remediation to remove the risk level. The vulnerability assessment process is shown in Fig. 2.8.

In addition to this, the difference between penetration testing and vulnerability assessment is illustrated in Table 2.1.

2.5.2 Types of Penetration Testing

There are different types of penetration testing, which are categorized according to organization requirements. This is also named pen testing.

- Black box penetration testing
- White box penetration testing
- Gray box penetration testing.

Black Box Penetration Testing

The black box penetration testing, tester testing the system without having any idea about the procedure. The tester was gathering the target system information

Table 2.1 Comparison of penetration testing and vulnerability assessment

Penetration testing	Vulnerability assessment
The scope of the attack is determined very clearly	Resources and assets of a given system are analyzed clearly
Test sensitive data	Each resource is analyzed, and threats are discovered
Performing target information gathering	Quantifiable values are allocated to the available resources
Final reports are provided when it cleans the system	Mitigate the vulnerabilities of resource
Performing environment review and document analysis	Comprehensive analysis should perform in the target system
Ideal for network architecture and physical environment	Ideal for the laboratory environment
It's for critical real-time system	It's for non-critical systems

while performing the testing process. Simply says that the tester checks the system's outcome and does not worry about how the result is achieved.

Advantage of Black Box Penetration Testing

- It does not require any specific language knowledge; also, the testers are not experts
- The tester verifies the system specifications and contradictions
- Testing is performed according to the user's perspective, not the designer.

The disadvantage of black box penetration testing

- Test cases design is more difficult
- It does not have everything
- It is not worth it when the designer is already analyzing the test case (Fig. 2.9).

White Box Penetration Testing

In the white box penetration testing, the tester delivered complete system information like source code, IP address, schema, OS details, etc. This testing process examines the system's internal sources for determining the simulation attacks. In addition to this, the testing process covers data flow, loop testing, path testing, and code coverage. This process is also called the open box testing, clear box, and glass box.

Advantage of White Box Penetration Testing

- It exercised complete independent paths in the modules
- This testing process checking and verifying the entire logical decisions
- Discovering the syntax error and typographical errors

Fig. 2.9 Types of
penetration testing

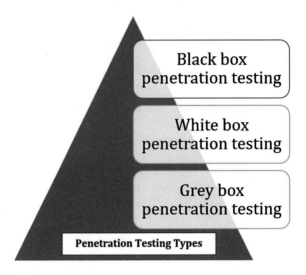

- Identifying the logical flow and actual execution by examining the designing errors.

Gray Box Penetration Testing

The tester is providing limited or partial information about the system or internal program.

Advantage of Gray Box Penetration Testing

- It is an unbiased and non-intrusive process. Therefore, the tester no needs to access the source code.
- Least risk of personal conflicts because of the exact difference between tester and developer provided.
- There is no need to provide internal details about operations and program functions.

2.5.3 Penetration Testing—Manual and Automated

The manual and automated penetration testing performed the same purpose, but they differ in the way of conducting. Machines do the manual penetration testing process completed by experts (human) and automated penetration testing.

Manual Penetration Testing

In this, testing human experts are continuously testing the programs or software and the risk, expert identifies vulnerabilities. It consists of several steps such as data

Fig. 2.10 Manual
penetration testing

collection, vulnerability assessment, actual exploit, and report preparations, illustrated in Fig. 2.10. The data collection process plays an essential role in testing, which can be done using tool services or manually. The tool gathers several information such as database version, table name, database, hardware, software detail, and other third-party plugins. The tester processes the collected data to assess the vulnerability factors and security weaknesses. Then, the tester launches the attack on the system to minimize risk. Finally, a report has been generated for protecting the target system from attacks.

This manual penetration testing is categorized in two ways such as,

- Focused manual penetration testing

  - It is focused on specific risks and vulnerabilities that are only performed by human experts. The automatic system does not perform this kind of testing.

- Comprehensive manual penetration testing

  - This testing focuses on the entire system to identify vulnerability and risk after identifying the vulnerabilities sorted according to the risk level. From the risk severity, this testing further analyzing whether the structure is affected by attacks or not.

Automated Penetration Testing

Automatic penetration testing works faster, reliable, easy, and efficient to automatically test risk and vulnerability. During the testing process, the system does not require any experts; instead, the testing is performed using that field's least knowledge. Several automatic penetration testing tools, such as OpenVAS, Nessus, back

Table 2.2 Comparison between manual and automated penetration testing

Manual penetration testing	Automated penetration testing
Different tools must be required to perform testing	Integrated tools are required from the outside of anything
Experts are needed to perform the testing process	Learners can perform the test because it is automated
Results are varied from one test to another test	Fixed results
The tester should clean memory	It does not need any cleaning process
Time-consuming and exhaustive	Fast and efficient
Experts are analyzing the target system situations, and security is established according to that	Machines do not analyze the situation
Experts run multiple testing	It does not perform multiple testing
It is more reliable during the critical condition	Not reliable

tract, are available to determine penetration testing excellence. Along with this discussion, the difference between manual penetration and automated penetration testing is described in Table 2.2.

2.5.4 Penetration Testing Tools

As discussed earlier, penetration testing is analyzing the risk and vulnerabilities of the collected information. The respective vulnerability exploits and report preparation are performed effectively. This process is achieved via several tools; hence, the tool's features must be learned to improve the testing process. Here, few penetration testing tools are illustrated in Table 2.3 for understanding the details of the tools.

2.5.5 Infrastructure Penetration Testing

In a business environment, computer systems and networks played a considerable role in performing the tasks. The devices and systems are interconnected, which may lead to damage at any time of the business process. Hence, the system and associative components must be protected from damage. For this purpose, infrastructure penetration testing is used to protect the internal/external devices, internet networking, virtualization, and cloud from vulnerabilities. There is the possibility of hacking the system infrastructure by the attacker. Therefore, it must be safe from security-related issues. Different types of infrastructure penetration testing are shown in Fig. 2.11

Table 2.3 Penetration testing tools

Name of the tool	Purpose of the tools	Portability	Cost
Nmap	Port scanning, network scanning, and OS detection	Linux, FreeBSD, Sun, OpenBSD, X-HP, NetBSD, IRIX, Windows Mac, etc.	Free
Hping	Remote OC fingerprinting, port scanning	FreeBSD, Linux, OpenBSD	Free
P0f	OS fingerprint, firewall detection	Linux, FreeBSD, Solaris, OpenBSD, AIX, Windows, NetBSD	Free
Superscan	Detects open ports (UDP/TCP) to determine which services are running on that port Queries running that include hostname, ping, whois, etc.	Windows 2000/XP/Vista/7	Free
Xprobe	Remote active OS fingerprinting, TCP fingerprinting, and port scanning	Linux	Free
Nessus	Vulnerabilities are detected that allow remote crackers to access sensitive data	Mac OX, FreeBSD, Win 32	Free
Iss scanner	Network vulnerabilities detection	Windows 2000, windows server 2003, windows XP professional with SP1a	Trial version Free
Hit print	Web server fingerprinting, detect web-enabled devices	Linux, FreeBSD, Mac OS X, Win32	Free
GFI LanGuard	Network vulnerability detection	Windows Server 2003/2008, Windows 2000 professional, Business/XP, Server 2000/2003/2008	Trial version Free
Brutus	Ftp, Telnet, password cracker	Windows 9x/NT/2000	Free
Metasploit framework	Create and execute exploit code against a remote target, computer system vulnerability testing	Unix and Windows entire version	Free
Shadow security scanner	Network vulnerability detection, LDAP server	Windows	Trial version Free

Fig. 2.11 Types of infrastructure penetration testing

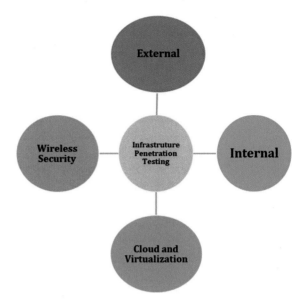

- External infrastructure penetration testing
- Internal infrastructure penetration testing
- Cloud and virtualization penetration testing
- Wireless security penetration testing.

External Infrastructure Testing

This kind of penetration testing continuously testing the external infrastructure against the hacker's perspective. Whether the network infrastructure is easily accessible via the internet or not, during the testing, a few attacks are continuously applied to identify the security flaws in external infrastructure. This process suggests a few guidelines for fixing issues, quickly identifying the information leaks, improving overall business productivity, highlighting the security risk, and successfully identifying configuration misuse.

Internal Infrastructure Penetration Testing

This penetration testing identifies the minor internal security flaws and fraud actions placed in large organizations. The testers are quickly identifying security possibilities, and respective employees are chosen to resolve this issue. This process identifies how an internal attacker takes advantage of minor security flaws, recognizes the potential business risk, maximizes internal infrastructure security, and provides detailed reports against the security exposures.

Cloud and Virtualization Penetration Testing

This testing process examines the attacker's activities in the cloud environment because most businesses use a third-party server to store the information. Hence, the cloud facilities and respective infrastructure must be predicted against the risk and vulnerability factors. This process determines the risk level with flaws, assists the action plan and guidelines to resolve the issues, protects the overall system, and provides possible solutions to the outliner security flaws.

Wireless Security Penetration Testing

The attackers are quickly hacking the system information from a remote location. The wireless security penetration process identifies the potential risks involved in wireless technologies, providing the action guidelines to recovering from external threats and improving the system's overall security.

2.5.6 Penetration Testing—Tester

The penetration tester role in an organization is critical because most critical information is difficult to protect from attacks. Therefore, here tester qualification, responsibilities, and experience are discussed, which are more useful for selecting the tester to perform penetration testing.

Penetration Tester Qualification

Qualification of penetration testing is more critical, while the tester performs the testing in a specific program or system. The testers must be qualified internally or externally before performing the penetration test. The testers are independent of management and organization. Before selecting the tester, they must complete particular certification courses that are listed as follows.

- Certified Ethical Hacker (CEH)
- CREST penetration testing certification
- Offensive Security Certified Professional (OSCP)
- Communication Electronic Security Group (CESG) IT health check service certification
- Global Information Assurance Certification (GIAC).

In addition to this certification, testers are selected based on experience. Then, the knowledge of the tester is recognized by the following questions.

- Penetration tester year of experience?
- Is the tester working independently from the organization?
- How many companies he/she worked as a penetration tester?
- Has he performed penetration testing for any organization that has a similar size and scope to yours?
- What type of experience does the penetration tester have?
- Asking references from other customers for whom he worked?

Along with these questions, testers handling complex situations, client requirements, and tester capability are checked via their earlier project. Based on the above criteria, the penetration tester has been selected, and they have the following roles.

- Identifying the inefficient allocation tools and technologies
- Internal system security should be testing
- Pinpoint exposures to securing the critical data
- Discovering the vulnerabilities and risks throughout the infrastructure
- Reporting and prioritizing the remediation recommendations that help to protect the data from hackers.

2.5.7   Penetration Testing—Report Writing

Report writing is an art that has been learned separately. Therefore, the tester does not know about report writing. The report consists of methodology, proper report content explanations, procedures, an example of a testing report, and the tester's personal experience. Completing the report writing has been shared with the senior management, staff, and technical groups. The report helps to recover the organization from future arises risks. The report consists of four stages: report planning, information collection, first draft writing, review, and finalization. The steps of the report writing are shown in Fig. 2.12.

Fig. 2.12 Stages of report writing

Report Planning

The report planning starts with penetration testing, like why the testing is importantly conducted and testing benefits. Then, time taken for the testing should be included. The primary subjects included in the report planning are listed as follows,

- Objectives—purpose of the pen testing
- Time—inclusion time, penetration testing scope on a specific time
- Target audience—includes the information technology manager, information security manager, technical team, and chief information security officers
- Report classification—classifies the report according to confidentiality like application information, IP address, threats, and vulnerabilities
- Report distribution—The number of copies should be included with the receiver's name.

Information Collection

As discussed earlier, penetration testing is more critical, and it is a lengthy process. Hence, the detailed collected information must be included in every stage of testing. In addition to this, tools, scanning results, systems, finding details, and vulnerability assessments must be included in the information collection process.

First Draft Writing

The first draft has been writing when the tester ready with their tools and information. The first draft writing includes entire activities, details, experience, and processes.

Review and Finalization

After drafting the report, it was reviewed by the drafter and then transmitted to the seniors to finalize the report. According to the above discussion, the sample report executive summary is described as follows.

Executive summary

- *Scope of the work*
- *Project objective*
- *Assumption*
- *Timeline*
- *Summary of findings*
- *Summary of recommendations.*

Methodology

- *Planning*
- *Exploitation*
- *Reporting.*

Detail Findings

- *System information details*
- *Windows server information.*

Reference

- *Appendix.*

2.5.8   Penetration Testing—Ethical Hacking

The growth and utilization of the internet increase because most public and private works depend on the internet. The organization and government work, plans, and operations are performed via the internet. Even though internet usage makes life flexible and accessible, criminal hackers are presented to hack the information. The hackers are creating cyberattacks and damage sensitive information. Hence, the data should be protected from the hacker by using the ethical hacker concept. The ethical hacker with several trustworthiness skills needs to identify risks, vulnerabilities, and maintaining the data confidential. They should know about hardware, networking, computer programming, and patience to complete penetration testing. Here, the difference between penetration testing and ethical hacking is described to understanding the concepts (Table 2.4).

Therefore, penetration testing effectively protects the data from threats and identifies the new weak points. Even though penetration testing is insufficient to safeguard security, it requires a new security concept to protect the system effectively. Therefore, the ethical hacking concept is utilized to identify computer systems and business attacks.

Table 2.4 Penetration testing versus ethical hacking

Penetration testing	Ethical hacking
Focus only to secure the security system	Penetration testing and the comprehensive term is one of the main features
Tester having comprehensive knowledge of the specific area to conduct the pen testing	Requires a comprehensive knowledge of software programming
Testers do not need to have report writing skills	Requires experts to report writing
Pentest has been performed by any tester having few inputs	Subject expert needs, and the person must complete an ethical hacking certification course
Paperwork is minimum	Detailed paperwork is need
Less time	Requires more time to perform ethical hacking
Does not requiring the accessibility of computer system and infrastructure	Requiring the entire system and infrastructure accessible permission

2.5.9 Penetration Testing Limitations

- Time limitation
- Scope limitation
- Limitation of access
- Limitation of methods End of the skill set of tester
- Limitation of known exploits
- Limitation to experiments.

Summary

This section explained the introduction to hacking and its basics, with the classification of hackers described in detail. Then, the section gives a detailed explanation about ethical hacking and types of cybercrime, denial-of-service attacks. Finally, the section concluded with penetration testing and its phases of testing. The next section deals with hacking for dummies.

Preview

This chapter covers the security hacking and classification; then, it covers the hacking techniques and tools with ethical hacking and their phases. Overview of security testing and covers proper hacking tools. It gives a detailed overview of physical security and its objectives, vulnerabilities, and finally, it covers risk assessment and their countermeasures and protection technique.

Assignment Question

1. What are all the types of penetrating testing?
2. List the difference between penetrating testing versus vulnerable testing?
3. What do you mean by a cross-site scripting attack?
4. What do you mean by symmetric key cryptography?
5. List the cybersecurity attacks and hacking?

Multiple Choices Question

1. List one of the subsequent SQL rows?

   (A) Process by (B) Validation by (C) **Order by** (D) Cluster by
2. Does the purpose of using TRUNCATE AND DROP is for security?

   (A) True (B) **False**
3. List the following, which is not in an ACID property?

   (A) Consistency (B) Isolation (C) Durability (D) **Availability**
4. Symbol, which is used for terminating A SQL query?

   (A) Quote (B) Comma (C) Question mark (D) **Semicolon**
5. What are all the following commands in SQL which is used to select only one copy of each set of duplicate rows?

(A) SELECT UNIQUE (B) SELECT DIFFERENT (C) **SELECT DISTINCT**
(D) None of the above.

Answers

1. **Order by**
2. **False**
3. **Availability**
4. **Semicolon**
5. **SELECT DISTINCT.**

Summary Question

1. Explain in detail about penetration testing?
2. Which of the following Linux commands is used to clear all the current IP tables rules?
3. Illustrate the importance of cybercrime and its limitation?
4. Give a brief explanation about ethical hacking and types of cybercrime, denial-of-service attack?
5. List the difference between manual penetrating testing and automated penetration testing?

Chapter 3
Hacking for Dummies

3.1 Security Hacker

A security hacker looks at safety violations and intrusion methods in a computer system or network. Hackers may be motivated by various factors, including benefit, protest, information collection, challenge, relaxation, or the assessment of system vulnerabilities to improve future hacker defenses. A hacker is someone working on computer and network systems protection mechanisms. White hat is the name given to hackers of ethical computers who make practical use of hacking. White hats are an essential part of the security field of information. They operate under a code that recognizes that it is terrible for people to break into other computers. Identifying and exploiting security mechanisms and machine failure remains a critical practice that can be carried out ethically and lawfully.

3.2 Classifications of Hacker

- **Cracker**

Somebody capable of subverting security on the computer. This can be called a cracker for malicious purposes, to hold the gap between alternatives such as the "cracker" and "hackers" inside the legal culture of programmers and computer breaks. Unauthorized admission to a computer to commit another crime such as destruction of the information in that device is an "unfrocking offense" is called a cracker or cracking.

- **White hat**

The internet word "white hat" refers to a computer hacker or an expert in computer health specialized in penetration tests and other testing methodologies that ensure the protection of the information systems of an organization. When operating in a

B. S. Rawal et al., *Cybersecurity and Identity Access Management*,
https://doi.org/10.1007/978-981-19-2658-7_3

technology company that produces security tools, a white hacker breaks defense on a non-malicious basis to test their security system and perform penetration testing or vulnerabilities evaluation for a client.

- **Black hat**

A black hat hacker (or black hacker) is a hacker who exploits personal gain or intent in computer security. The stereotypical, illegitimate hackers frequently portrayed in popular culture are black hats and are "the embodiment of all that the public fears in a criminal machine." Black hat hackers break into protected networking systems for data destruction, alteration, theft, or networks' insurability to approved network users.

- **Gray hat**

A gray hacker is between a black hat and a hacker with a white hat. A gray hacker could browse the internet and hack into a computer system to notify the administrator of a security defect in their approach. Occasionally, gray hat hackers notice a network malfunction and publish information worldwide instead of a community of people. Although hackers of gray hats may be excessively hacking, unauthorized access to a device can be considered illegal and unethical.

- **Green hat**

A hacker's green hat is not necessarily Irish, but it may be. Instead, a green that defines hackers who want to know the market's tricks even though they have no technical expertise and schooling.

- **Red hat**

A red hat hacker may refer to an individual targeting a Linux system. However, red hats are considered watchful. Unlike white hats, red hats are attempting to disarm black hats, but both parties' tactics are somewhat different. Instead of handing the authorities a black hat, red hats will initiate violent attacks against them to steal computers and equipment, damaging black hats.

- **Blue hat**

Blue hats look like white hats; the Blue Hat Conference of Microsoft was set up to promote contact between hackers and business engineers. A hacker who wants revenge is a blue hat. Blue hat hackers are like green hats too, but revenge is just the motivation of blue hat hackers—they do not want to develop their hacking skills.

3.3 Common Hacking Tools

To execute a perfect hack, hackers use a variety of methods, for example:

- **Rootkits**

A rootkit is a device or package that allows the threatened player to access the power of an internet-connecting computer machine remotely. A rootkit is a device or package that allows the threatened player to access the power of an internet-connecting computer machine remotely. Sadly, hackers are now using this software to destabilize their legal operator or user control of an operating system. There are different ways to mount rootkits in the network of victims, of which social engineering and phishing are the most common. Once rootkits are installed, the hacker can secretly access and control the system and offer them the chance to steal critical data from the program.

- **Key loggers**

A key logger is a specially crafted device that logs or records every system-pressed key. Key loggers record any keystroke as they type it via the computer keyboard by securing the application programming interface (API). The registered file will be saved, including usernames, visit data, screenshots, open applications, etc. Credit card numbers, text messages, telephone numbers, passwords, and other information may be registered for key loggers—as long as these are written. Key loggers are commonly used to intercept confidential data from cybercriminals.

- **Vulnerability Scanner**

A vulnerability scanner classifies and identifies different network device vulnerabilities, computers, networks for connectivity, etc. This is one of the most public techniques used by ethical hackers to find and quickly repair hidden loopholes. In addition, black hat hackers can use vulnerability scanners to search the network for potentially vulnerable areas and use the device.

3.4 Common Hacking Techniques

- **SQL Injection Attack**

To use the data in a database, a Structured Query Language (SQL) is intended. SQL injection is a kind of cyberattack aimed at trick machine databases through SQL statements. This form of attack is carried out using a web interface that tries to issue SQL commands through a database for hacking usernames, passwords, and other databases. Websites and poorly designed applications are vulnerable to SQL attacks as these web-based applications contain user data fields susceptibility and easily hackable through coding manipulation.

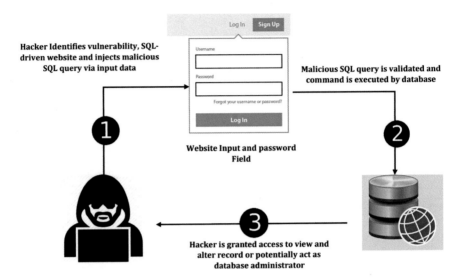

Hacker Identifies vulnerability, SQL-driven website and injects malicious SQL query via input data

Malicious SQL query is validated and command is executed by database

Website Input and password Field

Hacker is granted access to view and alter record or potentially act as database administrator

Fig. 3.1 SQL injection attack illustration

The SQL injection attack compromised bypassing authentication, data stealing, data corrupting, data deleting, running arbitrary code, and gaining root access to the system itself. The SQL injection attack is more dangerous because it accesses the credential business or individual information. Once the hacker hacks the sensitive information that is difficult to recover fully. The databases are attacked via the websites, but they are targeted directly. The sample SQL injection attack is illustrated in Fig. 3.1.

- **Distributed Denial of Service (DDoS)**

DDoS is a malicious form of attack which causes regular traffic, which floods network traffic into a server. It acts like a jam that obstructs the road and prevents daily traffic from reaching its destination. Devices that connect to the network easily are vulnerable to DDoS attacks. During this process, DDoS attacks need an attacker to carry out the attacks in the system. The attackers are remotely controlling via the group of bots that is called a botnet. The bot or zombie is nothing but the machine that is infected by malware. After creating the botnet, attackers continuously update, sending the instructions to the bot via remote control. Therefore, each bot responds to every request, while the botnet targets the victim's IP address. These attacks are led to creating the network overflow, and target servers are affected. The DDoS attack is illustrated in Fig. 3.2.

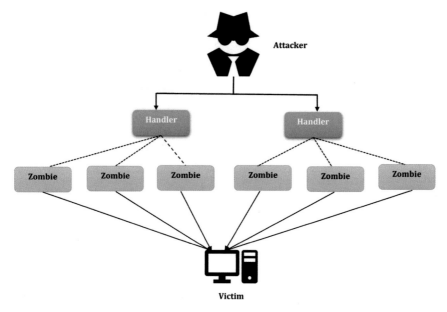

Fig. 3.2 DDoS attack illustration

3.4.1 Ethical Hacking

Proper handling is an accepted method of bypassing system protection to recognize potential network data breaches and threats. The system or network provider allows information security professionals to perform these exercises to test the system's defenses. Ethical hackers seek to investigate devices or network weak points that can be abused or damaged by malicious hackers. It collects and analyzes information to find ways to enhance system/network/application security. This enables them to increase the safety footprint so that they can better stand or distract from attacks.

Checking for Principal Vulnerabilities Involves But is Not Limited To

- Assaults by injection
- Security updates
- Sensitive data exhibition
- Infringement of protocol authentication
- Device or network elements that can be used as access points.

Objectives of Ethical Hacking

Hacking's goal is mostly untrue, i.e., crime or malicious intent to commit fraud or injure an individual, group, or organization, whether financially or reputably. It is achieved by stealing sensitive data or misappropriating funds or other monetary resources, disrupting companies, spreading inaccurate and malicious rumors, and

Fig. 3.3 Phases of ethical
hacking

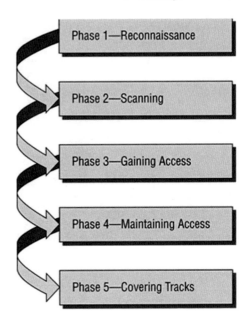

other socially damaging misleading news. Hackling is described as a type of internet or cybercrime that can be punished by law.

Passive and Active Surveillance

Passive recognition involves gathering information on a potential target without the knowledge of the target individual or organization. Passive identification may be as easy as watching a building decide when workers enter and exit the building, as shown in Fig. 3.3. Some appreciation, though, is achieved on the screen. When hackers look for information about a possible goal, they usually search the internet for information on an individual or company. This method is generally called information collection when gathering information on a TOE. Network sniffing is another form of passive identification that can provide useful information as secret server or networks, conventions, IP address scope, and other available computer or network services. Network traffic sniffing is close to building control; a hacker monitors the movement of data to see where these transactions happen and where traffic goes. Too many ethical hackers sniffing network traffic is a glowing lure. When the hacking devices are in place, all the information transmitted via communication networks is visible. Figure 3.3 shows the phases of ethical hacking.

- **Active identification**

It involves scanning the network to identify individual hosts, IP addresses, and network infrastructure. This approach involves a greater probability of detection than passive perception and is called doorknob rattling. Active recognition can provide a hacker with a safety sign, but the process often increases the possibility of being spotted or raising suspicion. Many software tools that carry out active recognition can be traced back to the machine that operates the software, increasing the risk of the hacker being found.

- **Scanning**

The scan allows the information obtained in the identification course to be extracted and used to search the network. Here, some of the tools are used for hackers during the scanning stage,

- SNMP sweepers
- The ping sweeps.
- The port scanner
- The web mappers
- Ping sweeps
- ICMP scanners
- The dialers
- The vulnerability scanner.

- **Gaining Access**

In the detection and scanning process, vulnerabilities are now exploited to gain access to the target device. The attack can be transmitted via local wired or wireless internet (LAN) wired to the target network, regional connectivity to a PC, the phone, and offline. For example, stack-based buffer overflows, service denial, and hijacking are included. Control in the hacker world is known as the network owners because the hacker has power over and can use the device; however, he wishes after a network has been hacked.

- **Access to be maintained**

A hacker has accessed the destination network; it needs this access to future manipulation and attacks. Often hackers harden the system by securing their unlimited permit to the Trojan, backdoor, and rootkit from other hackers and security personnel. When the hacker has the device, it can be used to conduct more attacks. In this case, the proprietary system is called a zombie program.

- **Covering Tracks**

When hackers have gained and kept access to the system, they cover their tracks to prevent safety personnel from detecting them, continue to use their system, remove evidence of hacking, or avoid legal action. Hackers are trying to remove all traces of an attack, such as log files or IDS.

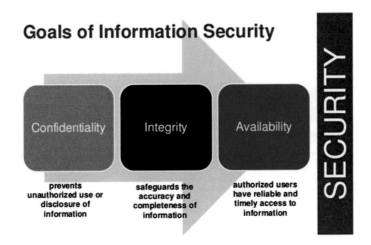

Fig. 3.4   Goals of information security

3.5   Developing Ethical Hacking Plan

The security checks aim to detect device risks and evaluate their possible weaknesses not to be disconnected or run. This helps to recognize all potential safety hazards on the device and helps developers overcome these issues by coding. Security testing is an arrangement of software testing that detects bugs, threats, and risks in the software and prevents disruptive attacks. Security tests are intended to identify any software system loopholes and weaknesses, leading to information loss, income, and reputation by the organization's employees or external staff.

As shown in Fig. 3.4, security checks are meant to recognize the system's risks and evaluate its possible weaknesses so that the device doesn't stop operating or is used. This helps identify all potential safety hazards on the machine and allows developers to overcome these issues by coding.

3.6   Types of Security Testing

Seven necessary forms of security checks are included in the manual for open-source security testing. The type of security testing is illustrated in Fig. 3.5.

Vulnerability Scanning: It is achieved with automated tools to search a device for established signatures of vulnerability. Special programs are developed to detect the weak points present in the applications or computer systems. The main drawback of vulnerability scanning is accidentally computer crashes. Then, the type of vulnerability scanning is illustrated in Fig. 3.6.

Vulnerability Scanning

Security Scanning

Penetration testing

Risk Assessment

Security Auditing

Posture Assessment

Ethical hacking

Fig. 3.5 Types of security testing

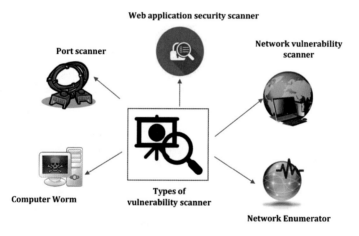

Fig. 3.6 Types of a vulnerability scanner

Security Scanning: This includes finding vulnerabilities in networks and processes and then offers solutions to reduce these risks. Scan for both manual and auto-mated scanning. This scan is possible. The security scanning process minimizes the mechanical examination of websites/applications/programs.

Penetration testing: It simulates a malicious hacking attack. During this test, a particular tool is tested to identify possible vulnerabilities to outermost hacking. Penetration testing has several phases, such as inspection, scanning, access getting, controlling access, and embracing tracks. This testing process is illustrated in Fig. 3.7.

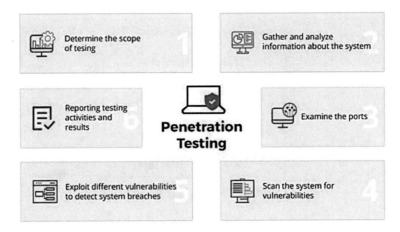

Fig. 3.7 Penetration testing

Threat Assessment: This involves analyzing the safety risks in the company. Hazards are classified as low, medium, and large. This test suggests risk management measures and actions. The threat assessment process depends on the process of formalization. The risk or threat assessment process is shown in Fig. 3.8.

Safety Auditing: It is an independent analysis of safety vulnerability in freeware and application software. The primary security verification type examines the system working conditions and configurations.

Posture Assessment: This incorporates security checks principles and threat assessments to appear to accompany overall safety status. This assessment includes security

Fig. 3.8 Risk assessment process

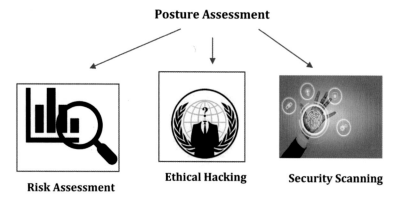

Fig. 3.9 Posture assessment

scanning, ethical hacking, and risk assessment. The phases involved in the posture assessment are shown in Fig. 3.9.

Ethical hacking: This exploits the operating systems of an organization. In comparison to malicious hackers who steal themselves, the aim is to reveal machine safety defects.

3.7 Ethical Hacking Tools

The primary goods used to secure and safeguard the various networks are ethical hacking methods and techniques. There are numerous ethics hacking tools from multiple sources available. Current technology experts use the latest proper hacking methods. These instruments help in the safety inquiry. In addition, we can see in detail the use of these tools.

- Ettercap
- Etherpeek
- Superscan
- QualysGuard
- WebInspect
- LC4
- LanGuard network security scanner
- Network stumble
- ToneLoc
- Netsparker
- Burp Suite
- John the Ripper
- Nmap
- Kismet

- THC-Scan
- Nessus
- Internet Scanner
- Nikto
- Ethereal
- Wireshark
- OpenVAS
- Acunetix
- Ethical hacking techniques.

Ethical hacking is the art of hacking white hats. The device can be checked, scanned, and monitored. Some of the methods for ethical hacking are the following. Say like

- Phishing
- Sniffing
- Social Engineering
- SQL injection
- Session hijacking
- Footprinting
- Enumeration
- Cryptography, and so on.

Such hacking strategies allow ethical hackers to protect systems and networks. Such methods are most effective if they are used by a skilled hacker to secure networks. These are more reliable and take advantage of the new system and device security updates.

3.8   Physical Security

The physical security of information-related asset storage systems, hard drives, computers, organizational machines, laptops, or servers can be described as safety and concern. The main focus of protection is on actual threats and crime such as unauthorized access, natural disasters such as fire and flooding, human disasters such as robbery. And thus, physical checks are required, such as locks, safety barriers, walls, and doors, which are not penetrable, continuous power supplies, and security staff to secure private and confidential information stored on servers.

- **Information Security Versus Physical Security**

The logical gap is in both words. The protection of information is generally about protecting data against unauthorized access, release, illegal use or alteration, recording, copying, or data destruction. The safety of data is a logical domain, while protection is based on the physical environment.

- **Objectives of Physical security**

  - Understand physical health requires
  - Identify threats to the physical security of information. Describe the main physical safety criteria for a facility site selection
  - Identify components of physical security surveillance
  - Understand the value of fire protection schemes
  - Describe the fire sensing and response components.

- **Factors depend on physical security vulnerability**

Regarding safety, any hacking may lead to success if the attacker has access to the building or data center of an organization that seeks physical safety vulnerability. The issue could be reduced in small enterprises and organizations. But, the following may be other factors that depend on physical safety vulnerabilities: How many places of employment, homes, or locations are there?

- Organization building size
- How many employees are employed?
- How many entries and exit points in a house are there?
- Reporting and other sensitive information for data centers.

- **Layers of Physical security**

Physical security depends on the model of layer protection like that of the security of information. The layer is installed and moves to an asset on the perimeter. These layers are:

- Deterring
- Delaying
- Detection
- Assessment
- Response.

3.9 Risk Assessment

All physical and cybercriminals are driven by the same reasons like money, the social schedule, etc. Intruders often seek to find ways to hack them. Those three terms—purpose, chance, and means—are thus given together to define the absolute risk of which is determined.

- **Countermeasures and Protection Technique**

Physical security is sensitive to security tests. From protection, other experts should be involved during the design, assessment, and updating phases. The truth is that security checks are always reactive. Besides the safety measures to be taken, it is necessary:

– Strong doors and keys. Solid keys
– Security cameras, especially around points of entry and exit
– Windowless data center walls
– Fences (with wire barbed or wire raster)
– The use and control of closed-circuit television (CCTV) or IP network cameras in real time must be done
– To identify unauthorized entries and warn a responsible person, an intrusion sensing device has to be added
– Know the different types of electromechanical and volumetric IDS systems
– To protect data from physical theft or injury, safety workers and guards must be used
– A simple type of biometric access system should be used (such as access to facial or fingerprint fingerprinting)
– Apply to physical protection with the Identification Card and Name Badge access control
– A range of lock systems, such as hand locks, computer-operated programmable locks, electronic locks, biometric locks (facial scanning and retina scans), may apply
– To communicate if there are incidents like fire detection, intrusion detection, vandalism, environmental hazards, or service disruption, alarm and alarm systems should be built to build infrastructure.

- **Adequate Physical security achieved in the organization**

In the event of physical protection, companies frequently use a copy–paste technique. You decide to do what other companies do to ensure safety. Their particularity and criticality are ignored. This approach may be useful for the organization's same kind and size, but not when they are different.

- **Full-fledged approach**

It might not be an acceptable option to make a significant budget for implementing complete physical protection from a fenced wall to gunned security guard and access control to drone monitoring unless the nuclear plant or military research facility is of a high safety nature.

Spending on physical security must be justified by a risk-based approach to rolling out security measures, as shown in Fig. 3.10. The introduction of physical protection would be like taking the medication without awareness of the illness without recognizing security threats and possible losses. A risk-based design physical security approach addresses mainly the high-priority threats. For example, a fireworks company should make it a top priority to minimize fire risk and not install a monitoring system.

Summary

This section explained security hacking, and its classification then gives a detailed explanation about hacking techniques and their tools. Then, the area explained ethical

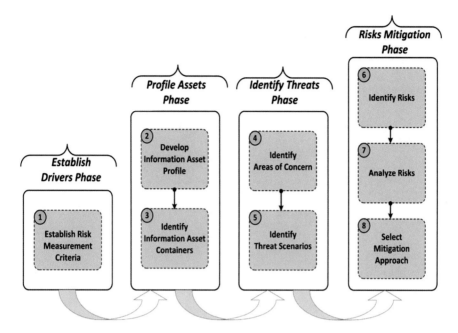

Fig. 3.10 Phases of physical security

hacking with their phases of ethical hacking and security testing. Finally, this section explained in detail the physical security and the risk assessment. The next section deals with the networking all-in-one for dummies.

Preview

This chapter covers the necessary networking and their types of topologies, operating systems, and their types. It covers hardware networking and what are the hardware components that connect to the network. It covers the network setup planning and design of the system, network security, and defense against computers sharing the network. Finally, it covers the problem solving and their methodology.

Assignment Questions

1. What are the advantages and disadvantages of hacking?
2. What is the difference between asymmetric and symmetric encryption?
3. How can you list the importance of ethical hacking tools?
4. What can an ethical hacker do?
5. Why is Python utilized for hacking?

Multiple Choice Questions

1. Most computer crimes are committed by …….

   (A) **Hackers** (B) International spices (C) Highly paid computer consultant (D) Web designers

2. Who gains illegal access to a computer system is known as……..

   (A) **Hacker** (B) Worm (C) Pirate (D) Thief
3. Certainties of the session ID need to check during login. True or False?

   (A) True (B) **False**
4. …….. is used to store the script in a vulnerable application?

   (A) **Persistent cross-site scripting** (B) ROM (C) Scripting (D) None of the above
5. ………OS that comes from the mobile manufacturer.

   (A) **Stock ROM** (B) New ROM (C) Old ROM (D) ROM

Answers

1. Hackers
2. Hacker
3. False
4. **Persistent cross-site scripting**
5. **Stock ROM**.

Summary Questions

1. List the importance of ethical hacking?
2. What are all the limitations used in security testing?
3. Explain in detail about physical security and risk assessment?
4. What are the common hacking tools used for classification?
5. Brief about developing an ethical hacking plan.

Chapter 4
Networking All-In-One for Dummies

4.1 Network Introduction

A computer system is a gathering of processors using a specific set of computer digital interconnection communication protocols to exchange information on or across network nodes. Interconnections are formed between nodes, a wide variety of innovations in the telecommunication network centered on the radio frequency, wired physically, optically, and wirelessly. This can be organized in several topologies for the network. It offers electronic communication across different technologies, interpersonal communication fax, online chat, text message, telephone calls, recordings, and video conferences. A computer network node can be defined as PCs, servers, networking hardware, or hosts for general purposes. The hostnames and network addresses are listed. Hostnames function as unforgettable node labels, after initial assignment, never modified. Network addresses are used to identify and locate nodes via protocols for communication such as the Internet Protocol.

4.2 Packet Network

Most modern networks use packet-mode transmission-based protocols. A network packet is a structured data unit that a packet-switched network carries. Packet network physical link technologies limit the packet size to a particular maximum unit before it is moved. A longer message is broken down. They are assembled to construct the original message after the packets have arrived. Packs are made up of two data types:

- User data
- Access management.

The control information provides data for the network's supply of user data, e.g., network addresses source and destination, error detection codes, and information sequencing. In packet headers and trailers, control information is usually contained

© The Author(s), under exclusive license to Springer Nature Singapore Pte Ltd. 2023
B. S. Rawal et al., *Cybersecurity and Identity Access Management*,
https://doi.org/10.1007/978-981-19-2658-7_4

data for the payload in the center shared between users rather than if the network was turned on. If a user does not send packets, the connection is possibly packed with other users' packets so that expenses can be spread if the link is not overused, with relatively little interference. In this case, the packet is queued and waiting for a free connection.

4.3 Topology of Network

The topological network is the organizational structure, form, or pattern contrary to their physical or geographical location; network hosts interconnect. Most network diagrams are typically arranged by topology. The topology of the network can affect performance, but reliability is often more critical. A single failure can occur with many technologies, including bus networks. Let the network crash. More interconnections are usually possible, the bigger the network is, but the more expensive the project.

4.3.1 Types of Topologies

– Topology of buses
– The topology of stars
– The topology of the ring
– The topology of the mesh
– The topology of the tree
– The topology of hybrid
– Topology relation connected.

- **Topology of buses**

All nodes of the bus network are linked to a specific network. This remains a topology on the layer of data link, even if modern versions of the physical layer use point-to-point connections. In this topology, network and computer devices are connected via a single cable; when the topology has two endpoints, it is called linear bus topology. The bus topology transmits the data in one direction and has a single cable.

Advantages and Disadvantages of Bus topology

The bus topology has several advantages such as cost-effective, utilized in small networks, requiring minimum cable, easy to understand, and easy to extend by joining two cables together. Even though it has several pitfalls, such as if the cable fails, the entire network fails, network traffic is high, and the whole network performance decreases, cable has a limited length and is slower than ring topology. Then the bus topology is illustrated in Fig. 4.1.

Fig. 4.1 Bus topology
structure

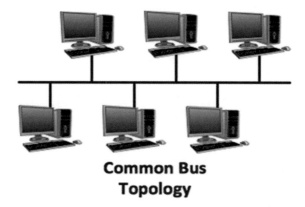

**Common Bus
Topology**

- **The Topology of Stars**

A unique central node links all nodes. This is the standard configuration of a wireless network, where the central wireless access point is connected to each wireless client. In the star topology, every node has its connection with the hub, which acts as a repeater to perform the data flow, and it has been used with twisted-pair such as coaxial cable or optical fiber.

Advantages and Disadvantages of Star topology

The star topology has several advantages: low network traffic, fast performance with few nodes; network works smoothly because only a specific node is affected, hub performance is upgraded easily, easy troubleshooting, easy setup, and modification. Although the star topology is having few pitfalls, such as expensive to use, high installation cost, the entire network is affected when the hub fails, and the performance entirely depends on the hub. According to the discussion, the star topology structure is shown in Fig. 4.2.

- **The topology of the Ring**

The left and right neighboring nodes of each node are linked to meet every node, and all nodes are connected. Crossing left or right nodes by other nodes, this topology was used by the Fiber Distributed Data Interface. In the ring topology, each node has two neighbors. According to the discussion, the ring topology structure is shown in Fig. 4.3.

Features of Ring topology

- This topology has several numbers of nodes that help to prevent data loss.
- Transmission is unidirectional, which is named a dual-ring topology.
- In dual-ring topology, the second ring back up the data if one ring fails during the data transmission process.
- Data transmitted in a sequential manner means bit by bit.

Fig. 4.2 Star topology
structure

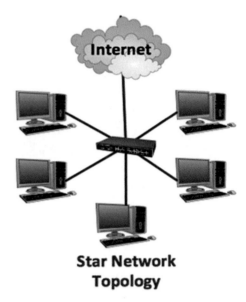

**Star Network
Topology**

Fig. 4.3 Ring topology
structure

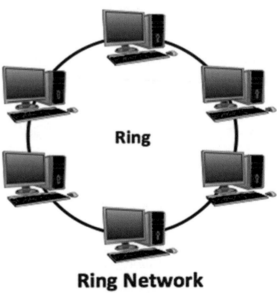

**Ring Network
Topology**

Advantages and Disadvantages of Ring topology

The network performance does not affect when it has high traffic, cheap to expand and install. Even though ring topology has specific pitfalls like difficult troubleshooting, network activities disturb when adding or deleting the computers, and whole network performance is affected by one computer failure.

- **The topology of the Mesh**

An arbitrary number of neighbors is connected to each node so that at least one node-to-one other traverse exists. In the mesh topology, information is transmitted in two types, such as routing and flooding.

In routing, mesh topology, nodes follow specific routing logic according to the network requirements. During this process, nodes are selected based on the link status, energy factor, broken links, node lifetime, and network lifetime because it helps avoid data loss. More ever, the nodes are selected to transmit the data from source to destination with minimum distance.

In flooding, mesh topology, data has been broadcasted to the entire nodes in the network. During this process, no routing logic is followed. The flooding-based created network is a robust but high network overload.

There are two types of mesh topologies, partial and full topology. In partial topology, systems are connected the same as the mesh, but few methods are connected to two or three devices. In full topology, each node is connected. As discussed above, the mesh topology is not flexible, robust, and fully connected. According to the discussion, the mesh topology structure is shown in Fig. 4.4.

Fig. 4.4 Mesh topology structure

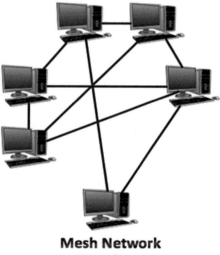

Mesh Network Topology

Advantages and Disadvantages of Mesh topology

The mesh topology has several advantages: robust, each connection having its own data load, easily fault identification, and managing security and privacy. This topology has few disadvantages, such as more cabling cost, requiring bulk wiring, and difficulty in installation and configuration.

- **The topology of the Tree**

In tree topology, nodes are connected like a hierarchy, in which root nodes are selected, and the remaining nodes are formed as a hierarchy. This topology is also named the hierarchical topology, in which the network has at least three levels of hierarchy. The tree topology is utilized in wide area network (WAN), and the network is ideal if workstations are located in the groups. According to the discussion, the mesh topology structure is shown in Fig. 4.5.

Advantages and Disadvantages of Mesh topology

The tree topologies have several advantages like easy node expansion, the performance of error detection is comfortable, an extension of bus and star topologies easily maintained and managed even though the network requires heavy cable, costly, difficult to manage when the network added more nodes, and if the central hub fails, then entire network fails.

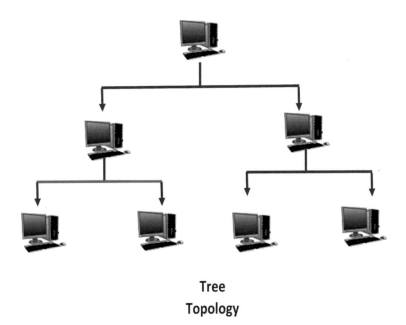

Tree

Topology

Fig. 4.5 Tree topology structure

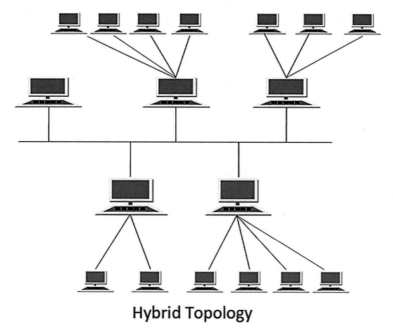

Hybrid Topology

Fig. 4.6 Hybrid topology structure

- **The topology of the Hybrid**

In this topology, two or more topologies are mixed, and new topologies are created. For example, if one system uses the star topology and the others utilize the ring topology, it is named the hybrid topology (star and ring topology). According to the discussion, the hybrid topology structure is shown in Fig. 4.6.

- **Topology relation connected**

The other nodes in the network will be linked to each node. Figure 4.7 shows the fully connected topology style.

4.4   Operating System for the Network

An OS that manages an operating network system capital of the network. Different operating system computers may be linked to the local area network. NOS-based software allows for many network devices to communicate with each other and to exchange resources. Hardware composition that usually uses a range of machines, a printer, is included in NOS. A local network server and file server that links them. NOS is then responsible for providing essential network services and multiple input requests enabling functionality in the multiuser environment. Figure 4.8 shows the operating system.

Fig. 4.7 Fully connected
topology structure

Fig. 4.8 Operating system

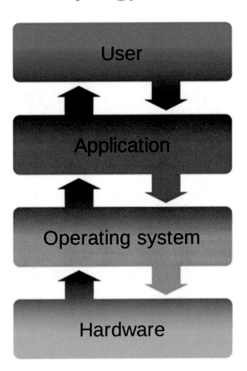

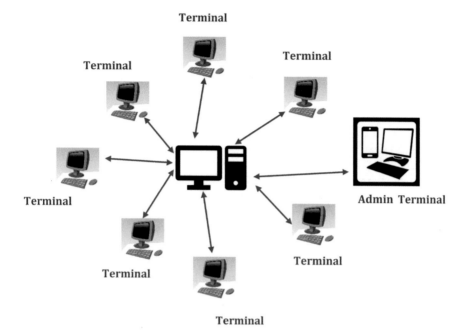

Fig. 4.9 Network operating system (NOS) diagram

The network operating system (NOS) was introduced in 1984; it is installed on the server-side network infrastructure. The discussed NOS gives much functionality like data managing, securities, applications, and other functions. The main objective of NOS is data sharing, printer access, and other devices. Then the structure of NOS is illustrated in Fig. 4.9.

- **Types of Network Operating system**

    - Peer-to-peer network operating system
    - Client/server network operating system

- **Peer-to-Peer Network operating systems** enable users to share saved common network resources. The network location is accessible. All devices are processed with similar functionality in this architecture. Peer-to-peer fits well for small and medium LANs and is less costly to deploy. In this NOS, users can share any type of files and resources via their workstation system, which is accessed by other computer systems.
- **Client/Server Network, operating system users,** access through a server to services. All features and applications are unified in this architecture on a file server to perform the client's actions. The most expensive way to incorporate the client/server and a lot of technical maintenance is needed. The client/server model has a benefit from unified network control, making it easier to introduce changes or improvements to technology.

- **Functions of Network Operating system**

  The network operating system (NOS) is having several functions that are listed out as follows:

  - The NOS helps to protect data, hardware components, and information from unauthorized users.
  - It permits the program testing routine.
  - It is able to manage the memory while loading the programs.
  - The NOS can detect real bugs and errors during job executions.
  - It establishes remote access to the client/server machine.
  - It is able to manage the sequence of processing jobs.
  - NOS permits entire users can creating user accounts.
  - It is able to manage the network resource configurations.
  - It permits the entire network communication services.
  - It monitors the network troubleshoots.

- **Characteristics of Network Operating system**

  - The network is supportable by a third party and several applications that can detect several machines' shared data and hardware.
  - Users can be able to manage the entire remote terminals because NOS supports operating systems' security features.
  - It was able to support internet functionality and able to manage their respective connections.
  - It can control the network security protocol, interconnected computer systems, and data backup process.
  - Users can back up their web services, database details and share their applications and printer.
  - Capable of performing clustering and the ability to support hardware, applications, and processors.

- **Advantages of Network Operating system**

  - Easy hardware configuration and stability
  - Establishes better protection
  - Terminals are accepted from different servers and various areas
  - It does not require any hardware equipment's
  - Current technologies are easily adaptable to the network environment.
  - Interoperability
  - Centralized servers do not require.

- **Disadvantages of Network Operating system**

  - The system does not have more protective
  - More expensive

- It does not have contains center storage space
- It requires frequent service and updating related requirements.

4.5 Hardware Networking

Networking equipment, known as network computers or computer networking equipment, electronic tools for connectivity and interaction needed computer networks between computers. They relay data in a computer network, in particular last user or data generation systems, as hosts end systems or tools for data terminals.

4.5.1 Range of the Network Hardware

Networking tools provide a wide variety of classifiable equipment as core network elements that connect other network components. The network's heart or boundary are hybrid components and hardware or software components that are generally located on different network connection points. Copper-based is currently the most common form of networked hardware—Ethernet adapter, available on most sophisticated computer systems. Wireless networking, particularly for portable and portable devices, has become increasingly popular. The data center infrastructure requires other networking devices in computers, network services, and content delivery devices.

4.5.2 Computer Components of the Necessary Hardware

- **Cables Network**

The broadcast of data from one system to another is network cables. Category 5 cable with RJ – 45 connectors is a widely used network cable.

- **Line Router**

A line router is a mechanism for connecting packets of data to different PC systems. Usually, they connect a high-speed internet connection to a network interface card from an organization. The RJ-45 ports are available for computers and other equipment using network cables you can connect to them.

- **Hubs, Switches, and Repeaters**

Repeaters, hubs, and switches connect network devices, and they can work as one unit. A repeater gets a signal before the hub is a regenerated redirect; that is how it is longer journeys to be walked. A hub is a repeater for multiport with input/output multiple ports so that every term has access. A switch gets port data and uses packet switches to solve them. The destination system forwards the data instead of transmitting it as a node to the actual destination.

- **Bridges**

A bridge links two different network sections. It moves packets to the intended network from the source network.

- **Gateways**

A gateway links entirely different networks executed on various TCP, UDP, and Ethernet. It is the input of the network and controls and quit point many networking links.

- **Tokens for application of the network**

Two types of network cards are internal and external cards.

4.6 Network Setup

- Link your computer. The internet connection to your home network is the router
- Access and lock the router application.
- Configure IP addressing and security.
- Establish sharing and monitoring.
- Configure user accounts.

Create a User Account on the Computer

When more than one person practices a similar machine, maybe you want every person to create a new account. You can build multiple user accounts using Windows. With Windows, you can create through user accounts. It is like accessing a specific device when every user logs on with a particular user account.

- Select the Start of the control panel. Click on the User Accounts. Add or remove connect in the browser. The Account Management dialog box will be shown.
- Click New Account Formation.
- Enter a name for your account and then choose the form of account to build.
- To close the Control Panel, click on Build Account.

Planning and Design of Computer Network

The following measures include computer network planning:

- **Identify the software you plan to use**: Networking of the machine can be significant, Enterprise Resource Management Systems, Instant messaging, email, and other services, internet telephony. The applications you want to use, such as the above, are essential to address. These are, in turn, used to estimate traffic requirements, software, and hardware.

- **Traffic specifications**: There are many contributors to computer traffic requirements. The following are a few things to consider:

  - Identifying and recording significant sources of traffic.
  - Traffic categorization client/server as a local, distributed peer to peer, end to host/server.
  - Each calculation device requirement for bandwidth
  - Service quality every program's requirement
  - Conditions for reliability.

- **Scalability requirements**: Scalability refers to network production that should be supported. Scalability is an essential consideration for the corporate network. Add users, applications, additional pages, and external network links should be provided.

- **Geographic considerations**: Take into consideration the required LAN and WAN links. WAN links can be used to connect offices separated by broad distance. In the same way, a LAN connection can connect building complexes within a compound. The LAN connections are typically high (10Mbps above), and the WAN connections are lower (64Kbps-2Mbps) bandwidths. LANs collapse onto a company's premises, while WANs typically, the Telecom is rented and maintained. WANs are costly and must be designed in terms of bandwidth and precisely planned to reduce energy use.

- **Availability**: When developing a network, the performance of a network needs to be considered. It takes time for a network to meet users; it is a critical design parameter. Sometimes, a clear connection exists both supply and affordability amount of redundancy required. Another essential element to remember in computing. The client's market loss is the quality requirements since the network has not been available for a specific time. An appropriate balance must be achieved to preserve profitability.

- **Accessibility and safety**: Protection and accessibility among the essential phases of architecture. A security plan that meets the necessary security requirements needs to be established.

  - A list of network services, such as File Transfer Protocol, site, and online mail that to be given.
  - Who controls the quality of these installations?
  - Why would people be taught plans and procedures on wellbeing?
  - Plan for the recovery event of a rights violation.

- **Price considerations**: In the case of LANs, hardware costs tend to be minimized. That reduces the cable costs, reduces the cost per terminal, and the cost of labor. The critical thing for WAN the goal is to reduce bandwidth usage to a minimum. The reason is that repeated bandwidth costs. Typically, machinery or labor costs are much higher. This gives effective equipment more weighting and bandwidth practical usage. Several cost-saving considerations are:

– Improve the performance with features like compression, voice behavior detection, etc., of WAN circuits.
– Using technologies like ATM, which allocate WAN bandwidth dynamically.
– Integrate voice as well as data circuits.
– Optimize or delete from the circuits used.

4.7 The Function of Prediction

In the network's planning and design phase, the traffic intensity predicted is estimated, and the network needs to sustain traffic load. If there is already a similar network, a network traffic measurement for the exact traffic charge can be calculated. If the network is incomplete, the network is to predict the planned transport speed; the planner will use telecommunications forecasting methods. There are several steps to the forecasting process: Fig. 4.10 shows the layout process.

• Defining an issue
• Define a problem
• Selection of the system of prediction
• Analysis/projection
• The findings are recorded and analyzed.

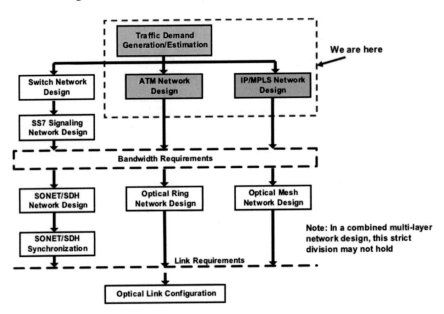

Fig. 4.10 Layout process

4.8 Network Security

Network security is a policy and practice of unauthorized access avoidance and monitoring, computer network assault, modification, or denial and services that are network accessible. Networks, like in a business, can be private, and others that may be publicly available. The network protection requires the authorization of data access in a network, and the network administrator is managed. Users select or are given an identity and password or another authentication. Data is allowing access within their jurisdiction to information and services. Network security includes a number of public and private computer networks, which are used in everyday work, transaction communications between corporations, government agencies, and individuals. Like in business, networks can be private, and some that could be freely available and some that could be freely available. The protection of the network includes organizations, businesses, and other forms of entities. It is the title; it protects both the network and activities to secure and supervise. The most expected and simple network is protected resources by assigning a special name and password. Figure 4.11 shows the network security.

- **Security of the information**

Employee actions can have a significant effect on organizational information security. Cultural principles may support growing divisions of the organization. Or work to ensure the confidentiality of information within an organization against effectiveness. The "complete pattern of" information security culture is conduct in an environment that helps secure all forms of communication.

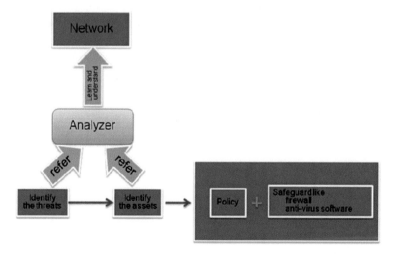

Fig. 4.11 Network security

- **Pre-assessment**: to evaluate knowledge, understanding employee health, and review the new security policies.
- **Strategic Planning**: to build a more flexible, there must be specific goals of Clusters to get it; people are supportive.
- **Operational planning**: developing a strong security culture management-buy-in, focused on internal contact, and knowledge of security and training the introduction of the information security framework will be carried out in four steps.
  - Management commitment
  - Communication with leaders of the company
  - Courses for all leaders of the company
  - Employee commitment.

- **Post-assessment**: to determine program and implementation performance and unresolved areas of concern to be found.

4.9   Defense Against Computers

A countermeasure is a tool, an operation in computer security procedure or technique reducing a hazard, danger, or removing or preventing an attack by minimizing. It could destroy it, or it could be detected and recorded to allow for remedial action. The following parts contain some common countermeasures; within this method, some methods include:

- The least privilege theory, where any element of the program has only the advantages for its purpose, they are essential and if an intruder. They have only limited access to the entire network, gain access to that portion.
- The automated theorem shows that essential device subsystems are correct.
- Code reviews and unit checks, modular approaches safer because there is no formal evidence of accuracy. Profound security where more than a subsystem is planned. The credibility of the program and the information it holds must be compromised to compromise.
- All flaws have been thoroughly exposed to ensure that when defects are found, the "Vulnerability Door" is as short as possible. Figure 4.12 shows the network security devices.

4.10   Sharing of the Network

Networking is a resource-sharing feature. Share files, records, directories, and media over a network via a connection to a laptop or other users/devices. This network can

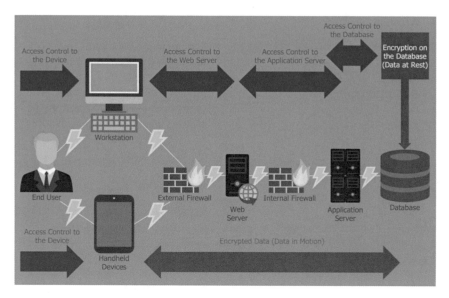

Fig. 4.12 Network security devices

share and exchange knowledge. This network will share and exchange knowledge. This network is produced by knowledge communication and trade.

4.10.1 Operation of Network Sharing

ICS routes TCP / IP packets from a small LAN to the internet. ICS offers NAT services that map local device IP addresses new exchanging device port numbers. Due to NAT's nature, IP addresses are not visible on the internet on the local computer. All packets that exit or join the LAN will be sent from or at the host ICS's external adapter IP address.

Usually, ICS can be used if several interface cards are available to host the computer built. In this case, ICS links to the internet-enabled for other applications on one network interface; this is known expressly as a private network. ICS is combined with UPnP from Windows XP enables remote ICS host discovery and control. It has a portion of the Quality of Packet Scheduler. An ICS client is connected to a reasonably rapid network. Windows can miscalculate the Optima through a slow connection to the internet TCP is given a window size dependent on the speed of the client-ICS host connection that could harm sender traffic. The TCP receive window size is specified as the ICS QoS component if the recipient were connected directly to the slow connection. ICS contains a Windows XP local DNS resolver for which name resolution is supplied. Both domestic network clients including networking tools are not Windows-based.

- **Wireless LAN**

A network of computers is a wireless network, and wireless communication connects to the network node. The way is through wireless networking homes, and telecom networks are used. And the costly process of inserting cables into a building avoids commercial installations or a link between different equipment locations. Wireless message is generally used to implement and manage telecommunications admin networks. This is carried out at the physical stage network structure of the OSI model.

- **Wireless networking styles**

  - **Open Wi-Fi**

Connecting devices in a relatively little area through wireless personal area networks (WPANs). It usually is beyond the control of a citizen. For instance, Bluetooth and invisible infrared light provide a WPAN for connecting a headset to a laptop. ZigBee can support WPAN applications. As equipment designers start, WPANs become commonplace to integrate Wi-Fi into a range of electronic consumer devices.

  - **Ad hoc network wireless**

Telephone ad hoc network or ad hoc mobile network wireless network. Every node transmits messages for the other nodes and routes every node. Ad hoc networks will "heal themselves," routing around a node that has lost power automatically. Ad hoc networks will "heal themselves," routing around a node that has lost power automatically. Diverse network layer protocols such as distance sequenced vector routing are needed to realize ad hoc mobile networks. Routing is based on associatively, ad hoc on-demand vector routing, and routing of complex sources.

  - **WLAN Home**

Metropolitan wireless networks are a form of WLAN connecting many WLANs. WiMAX is a wireless MAN type, and the IEEE 802.16 standard defines it.

  - **Mobility network**

A mobile network is a wireless network dispersed over soil areas known as cells; at least, one transceiver has been served for each. The base station or cell site is identified. The growing cell in a cell network uses a different character. Radio waves from all the neighboring cells to avoid interference. Both cells provide radio coverage over a broad geographical area when linked together. Both cells provide radio coverage over a general geographical location when linked together to connect anywhere in the network and with fixed transceivers and phones. Even if many transceivers pass over more than one cell during transmission through base stations. Figure 4.13 shows the link between two routers with one network.

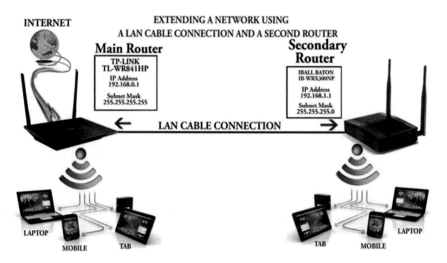

Fig. 4.13 In one network link, two routers

– Disorder Fixing Network

Problem-solving is a way of solving problems. They are applied regularly in a computer or device to repair faulty goods or processes and are routinely used in a computer or device to improve faulty goods or methods. It is a logical, systematic search and restarts the product or operation to solve the problem. Problem-solving is essential if symptoms are to be established; a removal cycle is the most probable cause of removing a problem's possible triggers. Finally, the solution must be checked to overcome problems restore to its working condition, the product, or the process.

4.11 Goals and Competences

Methodology goals and capabilities for problem-solving.

- Determine the problem
- Gather information
- If necessary, multiply the question
- Users of issue
- Symptoms classify
- Determine whether something has changed.

Figure 4.14 shows the steps to solve problem-solving that including faculty nodes, bottleneck, errors, changes, causes, and actions.

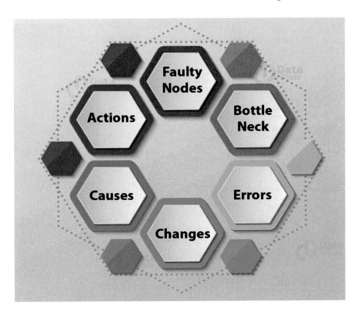

Fig. 4.14 Steps to solve problem-solving

Establish a probable cause hypothesis

- The obvious problem
- Take several approaches
- OSI model top-to-bottom-to-top
- Split and win.

Test cause determination theory.

- Upon confirmation of the idea, define the next steps to address the issue
- Re-establish new theories or escalate if the theory is not confirmed
- Create an action plan to determine the issue and identify possible consequences
- Implement or extend the solution as required
- Check the full functionality of the program and take preventive action if necessary
- Events acts, and outcomes of the paper.

4.11.1 Appearances of Problem-Solving

Typically, trouble-solving works with something that starts working unexpectedly, as its formerly functioning state is expected about ongoing behavior. The focus is often, therefore, on the current system or climate changes under which it operates. There is, however, a well-known theory that correlation means no causality. There is, however, a famous theory that correlation means no cause. Troubleshooting, therefore, involves logical thought instead of magic. Troubleshooting, therefore, involves logical thinking instead of magic. Problems can be solved by a systematic checklist.

Procedure for troubleshooting, flowchart, or table before the problem arises. In advance, it allows sufficient thinking to develop troubleshooting procedures on actions for troubleshooting and troubleshooting organization to the most effective method for problem-solving. Tables of troubleshooting can be computerized to increase user performance. It can either reply to further questions by the technician trouble-solving or decrease the list of solutions. The solution thinks it would solve the issue immediately. Such facilities are reimbursed if the technician goes a step further to document the solution after resolving the problem.

- **Network Problem-solving Flowchart**

To solve problems as effectively and painlessly as possible, some best practices are essential. It is important. Simplify the mechanism and prevent repetitive or wasteful efforts—flowchart for network computing.

- **Collect information**

To gather information first to support your end-users to ensure that the issue is clear to you. Gather sample knowledge from both the people's problems of the network and the network to treat the problem or duplicate it. Be cautious not to error on the root cause symptoms as it may be part of a larger issue that initially looks like the problem.

- **Changing charts**

Make sure you configure your event and security logs to send you details to help you overcome problems. A clear explanation of which items or things are reported in each log should be given, the date and time, and log source information.

- **Protection and access control**

Ensure there are no access or safety problems; check all access authorizations as they ought to be, and no one has altered a sensible aspect of the accident; they will not be able to reach a network. Search for all firewalls, anti-virus, and malware to ensure they function correctly and no security issues impact your users' capacity.

- **Employ a system of escalation**

Nothing is worse than that rather than going to the IT support desk and someone else, who then leads you to someone else who leads you to another. Have a simple structure for escalating who is responsible. The last person to contribute to the resolution in the chain is included. All your end-users should know who they can deal with a particular issue, so it is not time lost for five people to speak to who cannot solve the problem.

- **Using surveillance devices**

Problems can be overcome manually but can be solved. It takes time to take every step if you have a party knocking at your workplace or school door. It can be useful to send you frenzied emails to try to find the problem, fix it, and let us just fix it. In the company and industry circumstances, the easiest way to ensure that you get anything is by testing software information on the relevant network and is not missing to mention something important to avoid excessive risk exposure to the client.

The problem occurs when a diagram contains parts that differ from a product, many of which are available from your business. Suppose you record a troubleshooting protocol for, in a washing machine, the consumer solves a problem. For one model, they are troubleshooting, maybe 80% the same as for another. However, since individual components or features in one model might not be available in another, often, certain moves or tests will look otherwise in troubleshooting.

Even if your device will somehow render part of the diagram conditional, then conceal or show them according to the product ideally, troubleshoot documents that fix a specific issue. The work that occurred within a particular context will immediately be installed and presented to you as text or flowchart in a manner you like. After the knowledge is compiled automatically, the next move is to alter its depiction dynamically background or desires of the user. If this is a fast process to fix problems, it could be represented as is, which is as the technical writer's text. The method is complicated and takes different steps to make it simpler and quicker to understand every test's results. Or maybe both may be. Figure 4.15 shows the problem-solving flowchart.

The web automation framework transforms one of the application's interactive flowchart troubleshooting techniques and the interactive flowchart troubleshooting techniques. Since all product model information is included in the content repository, the diagram may be pressed, depending on the setup. A diagram check should pose a subject that explains how this search is done. If a new product model is now available and thus available, a new test or another adjustment must be made; it is just a question. To rerun the content and the diagram development process by clicking on a button. By the way, the code determines the optimum position automatically of diagram blocks to keep arrows from running over each other.

Summary

This section explained networking and its types of topology, and the functions of the operating system. It then gives a detailed explanation about hardware networking and the components that use for network and network setup planning and design. Finally, this section explained the network security and the defense against the computer and problem-solving methodology in detail. The next section deals with effective cybersecurity.

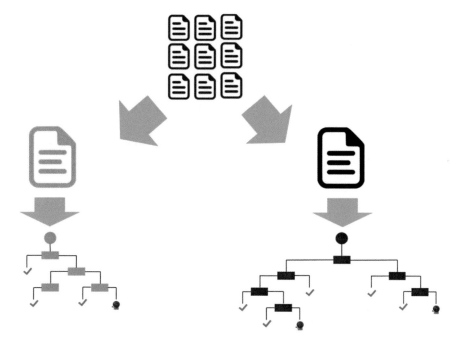

Fig. 4.15 Problem-solving flowchart

Preview

This chapter covers cyber-attack and vulnerability, security information, and the controls for defense. Then this chapter covers the network security governance framework and the evaluation of risk and the risk assessment, and then it covers the security evaluation aim and methodology approach security protection and the privacy policy, risk evaluation and penetration test, cyber danger and their facts, strikes cyber and the cyber protection of supply chain.

Assignment Question

1. Explain in detail network topology and its application?
2. List the types of topology and their advantages?
3. What is the importance of an operating system of the network?
4. What is all the range of network hardware used in network security?
5. List the defense mechanisms against computers?

Multiple Choices Question

1. ………is the standard interface of the serial data transmission

   (A) ASCII (B) RS232C (C) 3 (D) Centronics
2. How long is an IPv6 address……

   (A) 32 bits (B) 128 bits (C) 64 bits (D) 128 bits

3.   Two main types of access control lists

(A) IEEE (B) Extended (C) Standard (D) Specialized

(A) **2 and 3** (B) 1 and 2 (C) 3 and 4 (D) 1 and 4

4.   Most popular and comment Internet Protocol.

(A) HTML (B) NetBEUI (C) **TCP/IP** (D) IPX/SPX

5.   The term FTP means……..

(A) **File transfer protocol** (B) File transmission protocol (C) File transfer protocol (D) File transfer protection.

Answers

1.   RS232C
2.   128 bits
3.   2 and 4
4.   TCP/IC
5.   File Transfer Protocol.

Summary Question

1.   How do you set up a network interface in the system?
2.   What do you mean by the defense against computers based on network security?
3.   List the operation of network sharing and its application?
4.   What are all the goals and competence in networking?
5.   What do you mean by a function of prediction?

Chapter 5
Effective Cybersecurity

Cyber safety is information systems security and their hardware networks due to theft or damage. Software and electronic data, the services rendered by them from interruption or misdirection. This is increasingly important as computer systems become increasingly based. Wireless and wireless protocols like Bluetooth and Wi-Fi, and since "smart" devices have evolved like tablets, TVs, and various gadgets, It is the "Internet of Things." Because of its political and technical nature, cybersecurity is still one of today's biggest problems.

5.1 Attacks and Vulnerabilities

A limitation is a design, process, or internal control weakness; many known vulnerabilities are documented in the database of common exposures and vulnerabilities. At least one works for a utilizable flaw; there is an attack or "exploit." Vulnerability is hunted or abused regularly using automated software or custom scripts manually. It is necessary to safeguard a computer system to recognize the threats against security.

5.2 Community of Security of Information

The actions of employees can have a significant effect on organizational information security. Growing areas of the company may benefit from cultural principles that act efficiently or act for the performance safety of knowledge inside an organization. The "totality of trends" is the software protection community conduct in an agency relating to all forms' identity security.

- **Pre-assessment**: to identify information security awareness within the staff and to analyze the current safety policy.

© The Author(s), under exclusive license to Springer Nature Singapore Pte Ltd. 2023 87
B. S. Rawal et al., *Cybersecurity and Identity Access Management*,
https://doi.org/10.1007/978-981-19-2658-7_5

- **Strategic planning**: to develop a better consciousness programmed, there must be clear objectives. Clear objectives must be set. Clusters to do it, people are helpful.
- **Operational planning**: building a strong safety culture management-buy-in, focused on internal contact and awareness of security and training.
- **Development**: to use four stages to configure the culture of the safety of knowledge. They are the following:
  - Management engagement
  - Communication with members of the organization
  - Courses for all leaders of the company
  - Employee engagement.

- **Post-assessment**: to evaluate project performance and to identify unresolved areas of concern and implementation.

5.3 Cybersecurity Managed

Figure 5.1 shows cyber network security. SOC uses repetitive and vibration processes to systematically mitigate threats Info Sight analyzes risks in real time for all threat analyses carried out, individual cases are developed. Firewall system monitoring and control, Intrusion Detections/Appliances Avoidance, Host IPS, Active Directory, and Endpoint Systems are all available included in a cost-effective operation. Furthermore, all risks included incident-based reporting during the month, and it was

Fig. 5.1 Network cybersecurity

Detection
and
response

Prevention

An integrated portfolio that
enables orchestration

A focus on the fundamentals

A dedication to recruiting and retaining staff

An actual security strategy

Fig. 5.2 Network cybersecurity planning

analyzed and mitigated. The services for cyber defense and network management ensure your machine is secure 24/7 in a year.

- **Controls of defense**

Safeguards or countermeasures can be avoided, detect, mitigate, or reduce health hazards on land, documents, computer systems, or other equipment. It may classify them according to several criteria. Figure 5.2 shows the network cybersecurity planning.

- Preventive checks are intense before the incident to prevent an accident example. Unlawful intruders are being shut out.
- Detective searches are planned during the incident. Identify and describe an active event example. Intruding warning and alerting police or security guards.
- Corrective reviews are planned after the case limiting the degree to which the accident causes harm, for example, by restoring the organization's usual working state as quickly as possible.

  For starters, by their nature

- External checks example: clasps, gates set extinctions, and locks
- Command procedures such as response to incidents, oversight of supervision, health and training awareness
- User authentication technical controls example: logical access controls, tools for antivirus, firewalls;

– Controls of civil, administrative, or enforcement information privacy rules, procedures, and clauses.

5.4 Network Security Governance Framework

Creating an appropriate structure for knowledge governance is the best way to improve enforcement and risk management within the company. Ideally, enforcement should be defined within this framework. Blocks are representing the Regulation's essential features, global privacy procedures inside the system. Comprehensive collections of support instruments have been developed.

- **Impact of Privacy Analysis**

The anticipation is a central concept in the modern regulatory environment. Organizations should encourage privacy and comply with data security, ensuring privacy from the beginning of every new project from the earliest level. The risk is defined and controlled phase when new products and services are developed. Figure 5.3 shows the network security governance framework.

– **Crowd**: Type/volume of data collected in "as" limit collection needed for unique pre-authorized tasks
– **Approach**: Grant users' access to their tasks when they need to
– **Divide**: Data exchange access both inside and outside the organization
– **Apply**: Build systems that take into account user expectations
– **Cache**: Effective regulations on data protection
– **Clarification**: Where appropriate, using the aggregated, coded word, psychological or anonymous data

Fig. 5.3 Network security governance framework

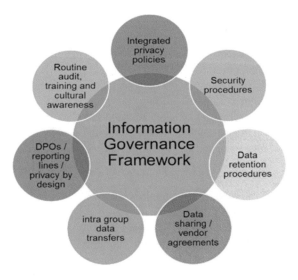

– **Regulation**: Build up to board level information management systems.

• **Evaluation of risk**

  – Identification and review of (future) possible events negatively impact individuals, properties, and the environment.
  – Make judgments on "risk tolerability" the basis of risk evaluation, "thus taking into account mitigating factors" (i.e., evaluation of risk).

A risk evaluation detects possible malfunctions, likelihood, consequences, and tolerances for these events. Quantitatively or qualitatively, the effects of this process can be conveyed. Risk assessment is an intrinsic aspect of a broader risk control strategy for helping to minimize future threats.

5.5 Risk Assessment of Structures

Risk evaluation on a much broader "program" scale may be carried out, or an example of a nuclear plant risk assessment. Systems can be defined as linear and (or complex) nonlinear, where they are linear. Systems with a change in the input are consistent and easy to understand, and unstable nonlinear processes when inputs are changed. As such, nonlinear/complex system risk assessments tend to be more difficult.

• **Complex assessment of risk**

The condition and threats are always present during an emergency response inherently less stable than planned. If the situation and the consequences are usually predictable, they should be covered adequately by standard operating procedures. It can be valid in some emergencies; adequate solutions to address the situation are planned and educated. The operator should handle danger in these circumstances without external assistance. And with the support of a contingency team that is trained and equipped for the short term. Or if an outsider party treats the situation, and they are not prepared for the situation, particularly what remains, but without any delay will deal with it. Figure 5.4 shows the target management system.

Present risk evaluation by the participants in these cases staff may advise appropriate risk management measures. In the changing situation, continuous risk evaluation is a corporate event to execute the control measures needed to ensure an appropriate safety level. The final step of integrated safety management is a complex risk evaluation system that can respond adequately under changing circumstances. This is focused on practice, preparation, and research, not just what went wrong, but successful debriefing, however, why and what went right to share it with others. Team leaders and staff responsible for risk management preparation degree. How statistics are transmitted and conveyed, the perception of profit and damage is influenced by both terms and numerically.

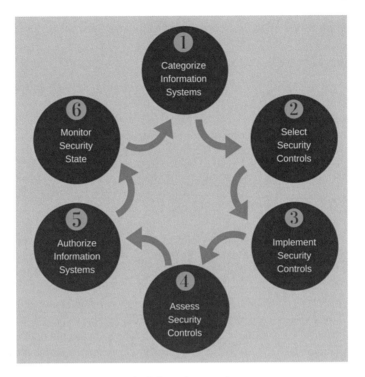

Fig. 5.4 Target management system for information security

5.6 Security Evaluation Aim

A health assessment's goal ensures that the safety measures required are incorporated into plan creation and execution. Any report documenting an adequately completed security evaluation would safety discrepancies between the design of a project and accepted company security policies. Assign the resources required to fix safety deficiencies or consider a risk based on an educated review of the risk/reward. Figure 5.5 shows the security attributes.

5.6.1 Methodological Approach

The next approach is presented as the conducting safety evaluation, effective means.

- Review of the situation and appropriate analysis
- Creating and reviewing security policy
- Analysis of the paper
- Review of threats
- Check for weakness

Fig. 5.5 Security evaluation attributes

- Review of data
- Report and review.

5.6.2 Checking for Protection

Safety checks are a method that exposes defects in the information network management systems protecting data and preserving as intended functionality. Owing to logical security check constraints, protection is passed testing does not mean there are no defects, or the system meets the safety criteria adequately—integrity, verification, preparation, permission, and non-reproduction. The existing safety standards are checked to depend on safety specifications systems introduced. Check health as a concept has various significances and can be fulfilled in a number of ways, as such, protection and importance by providing a simple working standard.

5.6.3 Protection and Privacy

A safety action that protects against information disclosure no other parties than the intended beneficiary shall mean the only way to guarantee protection.

- **Intelligence**
  - Data honesty applies to information security against modification by unauthorized parties that the device information is accurate.
  - Integrity mechanisms use the same basic concepts technologies, such as secrecy mechanisms, but typically to attach to communication knowledge, to form the basis of algorithmic analysis, instead of the entire communication encoding.
  - To verify that the correct information is passed from one application to another

- **Authentication**

The identity of an individual could be verified. Tracing an artifact's roots ensures that a commodity is what advertising and labeling claim to be or to guarantee a trustworthy product for a computer application.

- **Permission**

A determination that the applicant is approved to receive or run a service. A case in point is access management.

- **Offering services**

Data and coordination security, when expected, will be ready to use. Registered persons shall keep information accessible when they need it.

- **Non-repudiated**

    - Non-repudiation allows for information defense the parties sent and received a transmitted message claiming that the message was sent and received.
    - Non-repudiation is a way of ensuring the sender cannot dispute later that the message was sent and that the recipient cannot argue that the message has been received.

- **The taxonomic**

    Popular requirements for the delivery of safety tests:

- Exploration
- Check for weakness
- Evaluation of risk
- Assessment of health
- Check of Penetration
- Audit of health
- Examination of defense.

5.7   Evaluation of Risk

It enables you to instantly obtain a list of documented defects identifiable with a particular purpose for prioritizing remediation and in a short period through remediation tasks. The Vulnerability Evaluation tool simulates a generic automatically supplied low profile attack; however, the following main factors are not considered flaws in dynamic code management, unusual risk-causing configurations which cannot be detected user actions immediately, architectural or process problems are accountable for popular or new technology, on the other hand, etc.

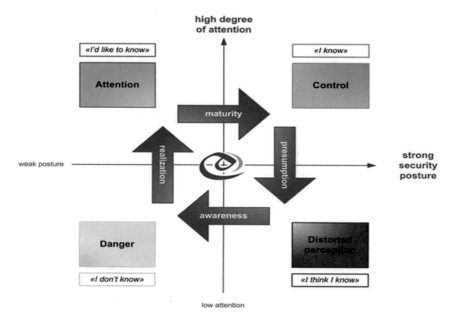

Fig. 5.6 Evaluation and penetration test of vulnerability

- **Tests for Penetration**

This reveals breach methods of an earlier reported and analyzes environmental threats and analyzes environmental threats the problem's source of remediation to deal with all the factors the attacker exploits testing the safety position of oneself by exposing experts to attacks. Check of penetration enables systemic, consistent, and repeatable assessment vectors of viability by multiple attacks used by an ethical hacker of the vulnerabilities assessment. It is possible to identify different vulnerabilities and show danger exposure, technically analyzing all application vulnerabilities complete or semiautomatic examination. Figure 5.6 shows the evaluation and penetration test.

5.8 Study of the Cyber Danger

The risk assessment refers to risk assessments related to the action or event in question. Risk identification applied to IT, projects, issues of security, and any other risk event the quantitative and qualitative basis is used for research. Risks are part of every business and enterprise project. The risk analysis will take place periodically to identify new potential threats and be updated. The strategic risk analysis reduces the risk of the probability and harm of the potential risk.

- **Risk analysis used by the company and organization**

  - Predict and reduce the impact of adverse events has resulted in harmful results.
  - Plan for the failure or loss of technology or equipment natural as well as human-caused adverse events.
  - To assess if a project is potential risks in the decision-making process when determining to drive the project forward.

- **Risk identification advantages**

Every business must consider the risks to connect the information systems protecting their IT properties effectively. Risk analysis will boost an organization in many respects, their health. The following are:

- Financial and organizational impact identifies, calculates, and compares the total effect of organization-related threats.
- It helps find information security vulnerabilities and identify the following steps to remove safety risks.
- It helps boost contact and information management decision-making processes and design cost-effective information management approaches, policies, and security procedures.
- It raises awareness of risks and safety measures for employees to understand, and during the risk analysis process, potential protection risks have financial impacts.

- **Steps in the process of risk analysis**

The fundamental steps followed by the process of risk analysis are:

- **Conduct an inquiry into risk evaluations**:

Get management and department details heads are critical to the process of risk evaluation. The risk evaluation study is to be started by documenting the department's specific risks or threats.

- **Determine the risks**:

Its phase is used for evaluating an IT structure or other aspects of a software-related risk identification agency, hardware, information, and IT staff. It detects possible adverse incidents in an organization like a human mistake, inundation, explosion, earthquakes, and flooding.

- **Risk analysis**:

After assessing and identifying risks, each risk occurring should be analyzed by the risk analysis process and identify the impacts associated with each risk. It determines how the objectives of an IT project can be affected.

- **Develop a plan for risk management**:

After the risk analysis that gives you an idea, these are essential assets, and what are likely to be threats? Negative impact on IT assets. A risk management plan to produce control would be to develop recommendations for prevention, transition, risk acceptance, or risk avoidance.

- **Implement the strategy to handle risks**:

The foremost objective of this step is to take action risks of analysis to delete or reduce. We can delete or reduce the risk from the first priority and resolve or mitigate every risk to prevent it from being a threat.

- **Overseeing the risks**:

The safety risk monitoring is responsible for this step for recognition, treatment daily, and risk control, which must be a vital part of the risk analysis phase.

5.9 Risk Evaluation Forms

- **Analysis of consistency risks**

The quality risk analysis method is a methodology for project management. This prioritizes project exposure by transferring the chance that happens as a danger case while impacting the risk occurrence consequences. Figure 5.7 shows the risk evaluation forms.

- Qualitative risk analysis aims to assess and measure the individually defined risk characteristics and then prioritize them based on the accepted features.
- The risk assessment assesses the likelihood every risk and effect on the project goals will occur.
- Danger categorization helps to weed out threats. Project risk exposure by increasing the likelihood and effects.

Fig. 5.7 Risk evaluation forms

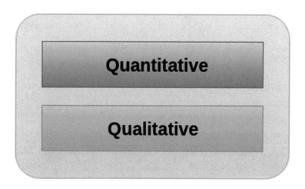

- **Study of quantitative risk**

  - The quantitative risk analysis process aims to provide the total impact numerical estimate damage on the project's goals.
  - It is used to measure the probability of success project targets and contingency fund calculation for time and cost ordinarily appropriate.
  - Quantitative analysis, especially for smaller projects, is not mandatory. Quantitative risk analysis can be used to quantify the overall project risk assessments that are of great importance.

5.10 Safety for Software Development

Security in software development is a mechanism of individuals and activities continuing, ensuring confidentiality, completeness, and software quality. The safety-conscious software development method benefits from stable applications where security is incorporated, and software protection is built. Safety is the most successful life cycle (SDLC) of software creation, particularly critical applications or critical information processing systems if designed and controlled throughout.

- **Fundamental concepts**

Computer protection is motivated by a variety of basic principles. Knowledge of these and how stakeholders should be technology applied is key to the security of users. These comprise:

- Defense from publication
- Protection against transition
- Protecting from loss
- Who makes the application?
- What are the applicants' rights and privileges?
- Capacity to produce factual facts
- Settings, sessions, and errors/exceptions management.

- **Checking for protection**

Popular safety testing attributes include authentication, authorization, anonymity, quality, honesty, prudence, and resilience. Safety testing is necessary to ensure the device prevents unauthorized users from accessing their information and resources; some client information is sent through the internet, collecting network appliances and servers. It gives ethical hackers plenty of space. With the rising concern and knowledge warfare practices that need fast responses, all appropriate is difficult to ensure information is provided to agencies and organizations to defend against the attacks of information warfare (Fig. 5.8).

Fig. 5.8 SDLC phase of planning

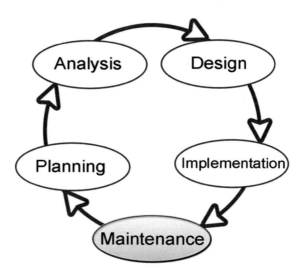

5.11 Strike Cyber

An attack attempts to reveal in computers and data networks unauthorized entry, impairment, kill, steal or gain allow unlicensed use of an asset; any kind of attack is a cyberattack—maneuver targeting operating systems, infrastructures, data networks, or computing facilities. An assailant is a person or process trying to access the information functions or other limited device areas without authorization or malicious purpose possibly. Cyberattacks can be part of cyber warfare or cyber-terrorism, depending on the context. Sovereign states may use a cyberattack, individuals, parties, organizations, or culture, and an anonymous source may come from it.

A cyberattack may steal, modify, or kill a specific target by installing spyware on a personal computer that may involve cyberattacks trying to kill the whole nation's infrastructure by hacking a susceptible device. Legal professionals aim to restrict accidents by using the word to cause physical harm and differentiate them from more familiar and more sophisticated and damaging cyberattacks.

5.12 Facts

- **Factor of spectacle**

The spectacularism variables calculate the real damage reached by an assault that leads to direct losses, and the malicious advertisement is obtained.

- **Factor of vulnerability**

  - How vulnerable vulnerability factor cyberattacks abuse an agency or government are created. Maintenance systems could operate organizations without old, more fragile servers than new systems.
  - A denial-of-service attack will make an organization vulnerable. And on a web page, the government institution may be defaced. A computer network attack corrupts the credibility or reliability of data.
  - Typically, by malicious code that modifies system logic which checks data and leads to performance errors.

- **Cyber protection supply chain**

The cybersecurity supply chain applies to actions that boost information protection in the supply chain. This branch of defense focuses on the supply chain on information security management requirements, informatics, applications, and network program threats such as cyber warfare, and ransomware. Theft of data and ongoing advanced threat typical information security supply chain practices for risk management cover just buying, disconnecting from trustworthy vendors from external networks, and notify users of risks and precautions they should take. The Principal Deputy Sub-Secretary for National Security, the US Department of Homeland Security Division, he knows that malware is present in instances. Imported products sold within the USA were found on electronic and computer systems.

- **Examples of cyber safety risks to the supply chain**

  - Computer network or device malware mounted on it has already been shipped.
  - Computer or hardware malware introduced technology and networking vulnerabilities.
  - Falsified hardware computer.

Figure 5.9 shows the anger control of the supply chain. Risk management supply chain (SCRM) is the implementation of daily and day-to-day management strategies; the supply chain has incredible risks based on ongoing risk evaluation to develop vulnerability and reliability ensure to negotiate with partners in or on your supply chain risks and confusion resulting from or impacting logistics operations or supply chain services.

Summary

This section explained about cyberattack and vulnerability, security information, and the controls for defense; then it gives a detailed explanation about network security governance framework and the evaluation of risk and the risk assessment with security evaluation aim and their methodology approach then it explains security protection and the privacy policy, risk evaluation and penetration test. Finally, this

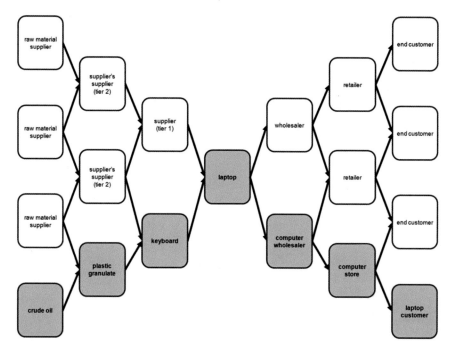

Fig. 5.9 Danger control of the supply chain

section explained the cyber danger and their facts, strikes cyber, and the supply chain's cyber safety.

Assignment Question

1. List the attacks and vulnerabilities in the cybersecurity process?
2. What do you mean by a community of security of information?
3. List the risk assessment structures used in the cyber-physical systems?
4. What are the methodological approaches in security evaluation?
5. Explain the protection and security policies?

Multiple Choices Question

1. In which of the following, a person is constantly followed/chased by another person or group of several people?

   (A) Phishing (B) Bulling (**C**) **Stalking** (D) Identity theft
2. ……. Is considered as the class of computer threats.

   (A) **DOS attack** (B) Phishing (C) Soliciting (D) Both A and C
3. is considered an unsolicited commercial email?

   (A) Virus (B) Malware (**C**) **Spam** (D) All of the above

4. Which of the following refers to violating the principle if a computer is no more accessible?

    (A) Access control (B) Confidentiality (C) **Availability** (D) All of the above
5. following used in the process of Wi-Fi hacking.

    (A) Aircracking (B) Wireshark (C) Norton (D) All the above

Answers

1. Stalking
2. DOS attack
3. Spam
4. Availability
5. Aircracking

Summary Question

1. What do you mean by strike cyber?
2. List the risk evaluation forms in cyber-physical systems?
3. How do you study the cyber danger and its mechanisms?
4. Illustrate network security governance framework
5. What do you mean by community of security of information and its advantages?

Chapter 6
Malware

The other important cybersecurity issue is malware. It is nothing but the software which ultimately affects the system, steals information, and creates confusion. The malware is classified into different types, such as ransomware, viruses, Trojans, spyware, etc. Respective malware tools like antivirus prevent this malware.

Hackers develop malware for earning money. Most of the time, the malware is intended to change the system against the security, protest the data, and even use it for government weapons.

6.1 What Does Malware Do?

The malware completely changes the files and has been classified into different types but is hardly comprehensive.

- Virus: It is an essential malware because it spreads quickly via the clean files and damages the other clean files. The viruses are damage the device's core functionality, also deleting the files. Usually, the virus malware files have appeared in terms of .exe (executable files).
- Trojans: This type of malware is embedded in the legitimate software that creates the backdoors and ultimately affects system security.
- Spyware: It is created for monitoring user activities. This kind of malware runs in the context and hacks user credential information like credit card details, passwords, bank information, etc.
- Worms: This malware damages the entire device, which is happened via the internet with network interfaces.
- Ransomware: This malware locks the system, hacks the user information, and creates threats to clear the complete report.

B. S. Rawal et al., *Cybersecurity and Identity Access Management*,
https://doi.org/10.1007/978-981-19-2658-7_6

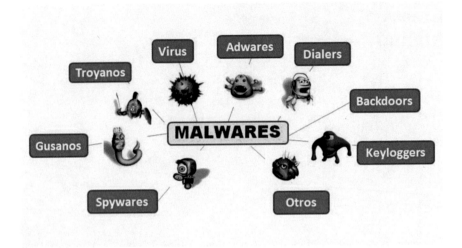

Fig. 6.1 Types of malware

- Adware: Though not continually malicious, aggressive advertising computer code will undermine your security simply to serve you ads—which might straightforwardly offer different malware in. Plus, let's face it: pop-ups square measure extremely annoying.
- Botnets: It is the networks of affected computer which is controlled and operated by the attackers.

  Then the graphical representation of malware is shown in Fig. 6.1.

6.2 How to Protect Against Malware?

The malware is more dangerous because it accesses our crucial and credential information without our knowledge. Therefore, malware must be prevented before it comes to affect our system. The chances of malware are reduced by using a few user behaviors, which are listed as follows.

- People do not have trust in new or strangers online. Most social engineering has fake profiles, abrupt alerts, emails, and curiosity offer with malware spreading files. Therefore, people do not trust and open unknown emails.
- Think before downloading the documents from the third-party server. Users must double-check the documents, files, and provider details to reduce the malware attack.
- Use the ad-blocker to be avoiding malvertising. The ad-blocker blocks unwanted advertisements, infected banners to access the device. This will be used to reduce the hacker's involvement in our system.

- Malware is presented anywhere; therefore, people should carefully utilize the browser. Several websites are poor security, which may spread the virus to infect the system.

Hence, healthy social sites and online habits are more helpful in eliminating malware involvement in the device. In addition to this, several antimalware software must be installed to detect the malware activities successfully.

6.3 Malware Analyzing Tools

The malware is analyzed and identified by utilizing different tools, which are classified into two types such as basic malware analyzing tools and dynamic malware exploring tools.

6.3.1 Basic Malware Exploring Tools

Several basis malware exploring tools are presented; a few tools such as PEiD, resource walker, dependency walker, file analyzer, and PEview are explained.

PEiD

It is the most common malware analyzing tool used to identify cryptos, packers, and compliers. Most hackers plan to write the malware code in terms of the package because it is difficult to detect while analyzing. The present PEiD malware detection tool can predict almost 479 signatures in PE files loaded via the text files. The small application of PEiD is shown in Fig. 6.2.

Dependency Walker

A dependency walker is another important malware analysis tool. It is a free malware detection application that is utilized in both 32-bit and 64-bit windows. The dependency walker helps to identify and analyze the imported and exported files and functions successfully. It also displays the dependency files and listed out the file path, machine type, version number, debug details, etc. The dependency walker malware application is illustrated in Fig. 6.3.

Resource Hacker

ResHacker or resource hacker also a free malware analyzing application. This tool helps to derive the resources from the windows binaries. The current version of ResHacker is 4.24 that was released in July 2015. It can derive, modify, and add several resources such as images, strings, version information, menus, dialogs, manifest resources, etc. The resource hacker malware analyzing tool is shown in Fig. 6.4.

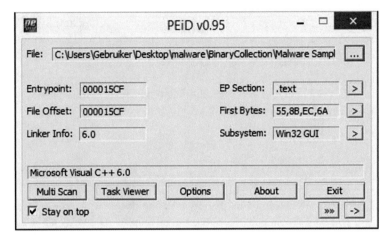

Fig. 6.2 PEiD malware analyzing tool

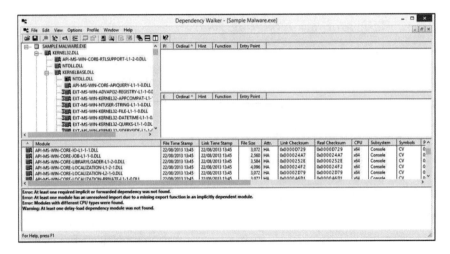

Fig. 6.3 Dependency walker malware analyzing tool

PE View

This malware analyzing tool is free and easy to utilize in applications. The PE view tools help analyze the files presented in a portable executable file and various file sections as shown in Fig. 6.5.

File Analyzer

The free malware analyzing tool helps to read the details from the PE headers and respective sections. The file analyzer provides a lot of information about the header files and features when compared to the PEview. The extracted features are saved

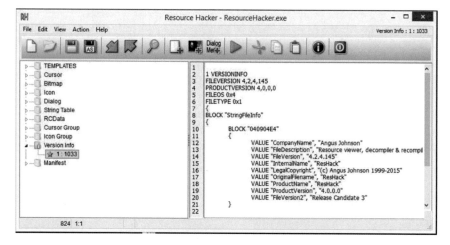

Fig. 6.4 Resource hacker malware analyzing tool

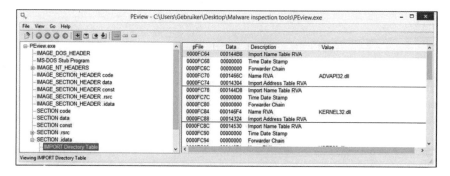

Fig. 6.5 PE view malware analyzing tool

in the virus tool that is forwarded to the virus tool analysis process to predict the malware related information. The representation of the file analyzer is illustrated in Fig. 6.6.

6.3.2 Dynamic Malware Analyzing Tools

A dynamic malware analyzing tool works after completing the basic or static malware analyzing process. This tool helps to examine the activities after the employment of malware viruses in the system. Several dynamic tools, such as process explorer, procmon, ApateDNS, Regshot, Wireshark, Netcat, and INetsim, perform the malware analysis process. These tools are more powerful because they quickly identified the type of malware in the fastest manner.

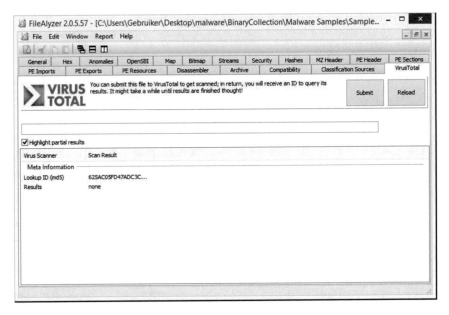

Fig. 6.6   File analyzer malware analyzing tool

Procmon

Process monitor is also named Procmon dynamic malware analysis tool, which Windows SysInternals created. This tool helps to monitor the windows process activities, file systems, and registry in real-time. This tool was developed by combining two legacy tools such as RegMon and FileMon. These tools help filter the entire data in the windows system and display the filtered or affected data Fig. 6.7.

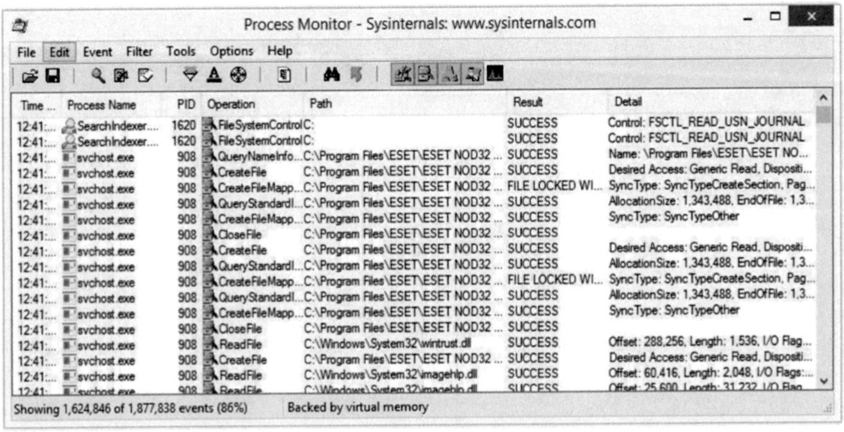

Fig. 6.7   ProcMon malware analyzing tool

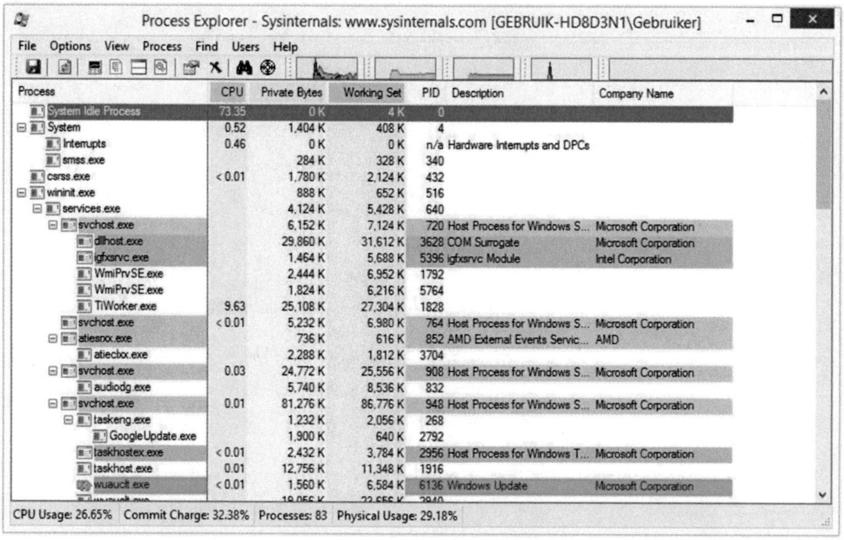
Process Explorer - Sysinternals: www.sysinternals.com [GEBRUIK-HD8D3N1\Gebruiker]
File Options View Process Find Users Help
Process CPU Private Bytes Working Set PID Description Company Name
System Idle Process 73.35 0 K 4 K 0
System 0.52 1,404 K 408 K 4
Interrupts 0.46 0 K 0 K n/a Hardware Interrupts and DPCs
smss.exe 284 K 328 K 340
csrss.exe < 0.01 1,780 K 2,124 K 432
wininit.exe 888 K 652 K 516
services.exe 4,124 K 5,428 K 640
svchost.exe 6,152 K 7,124 K 720 Host Process for Windows S... Microsoft Corporation
dllhost.exe 29,860 K 31,612 K 3628 COM Surrogate Microsoft Corporation
igfxsrvc.exe 1,464 K 5,688 K 5396 igfxsrvc Module Intel Corporation
WmiPrvSE.exe 2,444 K 6,952 K 1792
WmiPrvSE.exe 1,824 K 6,216 K 5764
TiWorker.exe 9.63 25,108 K 27,304 K 1828
svchost.exe < 0.01 5,232 K 6,980 K 764 Host Process for Windows S... Microsoft Corporation
atiesrxx.exe 736 K 616 K 852 AMD External Events Servic... AMD
atieclxx.exe 2,288 K 1,812 K 3704
svchost.exe 0.03 24,772 K 25,556 K 908 Host Process for Windows S... Microsoft Corporation
audiodg.exe 5,740 K 8,536 K 832
svchost.exe 0.01 81,276 K 86,776 K 948 Host Process for Windows S... Microsoft Corporation
taskeng.exe 1,232 K 2,056 K 268
GoogleUpdate.exe 1,900 K 640 K 2792
taskhostex.exe < 0.01 2,432 K 3,784 K 2956 Host Process for Windows T... Microsoft Corporation
taskhost.exe 0.01 12,756 K 11,348 K 1916
wuauclt.exe < 0.01 1,560 K 6,584 K 6136 Windows Update Microsoft Corporation
CPU Usage: 26.65% Commit Charge: 32.38% Processes: 83 Physical Usage: 29.18%

Fig. 6.8 Process explorer malware analyzing tool

Process Explorer

Process adventurer is also a free tool from Microsoft that should be running once acting dynamic malware analysis. Method adventurer is employed to watch the running methods and shows you which handles and DLL's area unit running and loaded for every process Fig. 6.8.

Regshot

It is one of the open-source dynamic malware analyzing tool which helps to screen the registry. During this process, the changes present in the registry have been screenshotted that are compared with the current registry state. This tool widely helps identify the registry changes after the execution of malware in the system Fig. 6.9.

Risk of Dynamic Malware Analysis Tools

As discussed above, dynamic malware tools can quickly predict malware but have several risk factors. The active analysis tool creates the network and system at risk because it is detected during the system execution. Therefore, any dynamic malware analysis tool should be run in the virtual machine or isolated networks to eliminate unnecessary errors. Although the dynamic tools work on virtual machines, there is no guarantee of the host machine's safety. By considering these risk factors, the above discussed active malware analyzing tools should be handled.

Fig. 6.9 Regshot malware analyzing tool

6.3.3 Difference Between Static and Dynamic Malware Analysis Tools

The developments of techniques interest cybercriminals to create different malware types, causing difficulties while performing malware detection. The malware detection and analysis process examine the origin's system, and its impacts and its functionalities are determined. Therefore, in cybersecurity, malware detection is one of the important vital challenges. Security analysts are interested in analyzing the malware details for a different purpose. Most of the analyzers doing their research to find the malware attacks and infected file information. Few professionals create the malware virus to understand the modules' functionality and working process, which helps handle cyberattacks. As discussed above, various malware analyzing tools are presented in both static and dynamic.

The static or fundamental malware analysis is used to analyzing the sample before executing the code. This process is performed with the signature verification process's help because it is a unique identification of the binary information. From the signature, components of the binary files are examined and determined using the cryptographic hash function. The binary malware file is uploaded to the dissembler during this process, and the machine-executable binary file code is changed into the assembly language code. This conversion process makes it easier for the analyzer to understand the binary file characteristics. From the assembly code, the malware-affected part and functionalities are easily identified. The malware analysts utilize various static analysis processes such as virus scanning, file fingerprinting, packer detection, debugging, and memory dumping.

The next one is dynamic malware analysis, which is different from the static malware analysis process because it detects the malware while running the code in the controlled environment. During this process, malware is executed in an isolated and closed environment for understanding the malware characteristics. Once the malware changes the module behavior and functionalities, execution is stopped to eliminate the infection from the files or modules. Therefore, the dynamic malware analysis does not miss any critical action and functionalities from the malware detection process. From the discussion, the difference between the static and dynamic malware analyzing tools is listed as follows.

- The dynamic analysis tool is worked according to the behavior process, whereas static analysis is signature-based.
- In dynamic analysis, malware codes are executed in a sandbox environment, whereas code is not running in the static analysis.
- In static analysis, the malware identification task is simple and easy to identify malware behavior. In dynamic analysis, the analyst determines the code functionalities, actions, and malware impacts in every phase.
- The static analysis is standard for all types of malware, whereas dynamic analysis is working according to behavior-based; therefore, it requires additional malware information.

The analysis clearly shows that malware analysis played a vital role in understanding malware infections and avoiding malware spreading into directories, files, etc. The discussed static and dynamic analyses were used to eliminate further attacks.

6.3.4 Warning Signs to Identify Malware Infection

In general, most people utilize antivirus to protect their systems from malware and virus infection. Hence cybercriminals are developed their ideas to affect the systems via malware and virus while running the security software. Several users think that security software eliminates unwanted activities; therefore, they fail to monitor the background actions. Here few warning signs are discussed to identify the malware infection.

- Notable decrease in system speed
- Applications consume more time to start or showing errors
- Often system crashes
- Maximum internet traffics
- Suspicious messages on social networks
- Difficult to access the control panel
- Alarm from false security programs
- Shortcut files in pen drive or systems
- Traffic redirections to various browsers
- The system is transmitting or receiving unusual messages.

6.4   Most Dangerous Malware of 2018

From Becker hospital's report, 91 patients' records are hacked by IT hackers, and 91 records are disclosure breaches. A portion of those incidences enclosed malware, a malignant software system meant to profit the cybercriminals. A system will finally end up tainted with malware through email phishing efforts, and a current Mimecast report discovered thirteen % of the messages contain hurtful substances like spam or malware.

Hackers will infuse malware into many devices with various final objectives, such as stealing info, manipulating the PC perform, or hijacking access controls. Also, hackers often convey malware assaults, so the merchandise goes undetected for AN extended stretch of your time. There are eight different malwares that critically affect the business process, and the malware is listed as follows.

- Ransomware
- Remote administration tools
- Adaptive malware
- Multipart malware
- Memory only malware
- Banking Trojan
- System-level malware
- Scripting malware.

6.4.1   Malware Detection Techniques Used by Antivirus

Antivirus is also named antimalware, which utilizes different analyzing techniques, technologies, and algorithms to identify malware types. Ten different malware detection techniques are being used to predict spyware, adware, viruses, Trojan, worms, horses, and other malware.

- Heuristic analysis
- File signature analysis
- Cloud analysis
- Behavioral analysis
- Host intrusion prevention systems
- Sandbox analysis
- Firewall
- Custom domain name systems and servers
- Web browser extensions/add-ons
- Web filtering and application control 2nd.

File Signature Analysis

It is the simplest and oldest technique used for analyzing malware. The file signature analysis consists of a database that includes a set of previously detected malware signatures. This database is named as the signature database or virus definition. With the signature's help, antivirus continuously scans the computer programs and files to match similar signature information. Suppose the program has identical signature information; it has been blocked and notified to the users.

Heuristic Analysis

It is the advanced version of the file signature analysis process because it protects the malware actions' files. The heuristic process utilizes the algorithms and techniques to predict the malicious or not. During this analyzes, program code is continuously examined, and various outcomes are expected. If the desired code exists in the signature database, then the entire program is blocked due to the malware's new variant.

Behavioral Analysis

This kind of examination helps to identify the infected programs which are not received from the virus definitions. This process examines the program behaviors and matches the malicious program. It includes the host intrusion prevention system and intrusion detection system for making the behavioral analysis process. Even though the behavioral analysis works effectively.

Cloud Analysis

Malware analysis is done via the cloud, which means antivirus vendor's servers. This analysis helps to predict the new kind of malware. During the investigation, if the antivirus detects the malicious program, then the signature is created and blocked from the other computer programs. One of the significant disadvantages of this analysis is continuously requiring active internet connections.

Sandbox Analysis

This type of analysis utilizes behavioral and virtualization analysis and detects the running program's malware actions. The sandboxing feature is also accustomed to running all of the files that the antivirus can neither whitelist nor blacklist. The files are dead in AN isolated section cut different loose files hold on the pc. So, running a go in a sandbox instrumentality offers you the most effective of each the worlds. If the file was malicious, it can't damage your pc because it had been dead in a very virtual environment/sandbox instrumentality. If it had been a legitimate program, you were ready to run it.

Host Intrusion Prevention System

It is used to detect malicious activities in the program, which is done by applying secure software. The system examines and monitors every program's activity and notifies the user if it has any malicious activities.

Web Filtering and Application Control

It helps to protect the systems from internet-based threats by blocking malicious browsers and websites. The application control is the process monitoring that helps monitor the program's behavior, functions, and activities while running the computer system program.

Web Browser Extensions/Add-Ons

The most significant way of getting malware infection is via web browsers. Due to the development of techniques and the internet's growth, most people utilize web browsers. Therefore, malware is entered into the computer via web browser extensions, plug-ins, add-ons, browser helper objects, etc.

Custom Domain Name System and Server

Most of the internet service providers introduce a new type of advertisement in the networks. These ads cannot block by an ad-blocker extension, which may cause the entire domain damage. Therefore, domain name system services are utilized to edit the host's files to secure the system from malicious activities.

Firewall

The effective security mechanism is a firewall because it was continuously monitoring the network connections (incoming and outcoming). From the monitoring report, malicious traffic and malicious applications are detected and blocked successfully. This process helps to manage sensitive data from the hacker. In addition to this, the system has been protecting from unauthorized access.

6.4.2   Tips to Prevents the System from Malware Actions

The malware has to be identified in the earlier stage; otherwise, the hacker can access several sensitive pieces of information. Therefore, here a few tips are listed out to prevent the system from malware actions.

- Usage of well protective antivirus
- Use of firewall
- Utilizing the on-demand malware scanner to perform the regular computer scan
- Continuously scan the removal devices during the usage
- Disable auto play and autorun
- Use safe websites for downloading software
- Manage the control of system startup
- Try to avoid spam opening
- Don't save documents, folders in libraries, desktop, and user folders
- Avoid using free Wi-Fi
- Continuously check Wi-Fi settings
- Keep software in up-to-date

- Run unknown programs in a virtual environment
- Enables security software and self-protection tools
- Use customized domain name server service provider
- Keep regular backups
- Keep bootable system repair disk and rescue disk.

Summary

This section explained the malware, types of malware, and malware's impact on computer systems. In addition to this, different malware analyzing tools, techniques, and types are discussed to eliminate the threats related to risk. Finally, this section explained the few malware protecting ideas and tools to avoid the system's unnecessary behaviors.

Assignment Questions

1. What does the malware do?
2. List the malware analyzing tools?
3. What do you mean by dependency malware analyzing tool?
4. How the resource hacker un-stabilize the system using their protocols?
5. What do you mean by PE view malware tool?

Multiple Choices Question

1. There are …….. types of computer viruses.

    (A) 1 (B) 5 (C) **10** (D) 11
2. List the following, which is not a type of virus?

    (A) **Trojans** (B) Boot Sector (C) Multipartite (D) Security
3. A computer ……… is a malicious code which self-replicates by copying itself to other programs.

    (A) **Virus** (B) Application (C) Worm (D) Program
4. Which of them is not an ideal way of spreading the virus?

    (A) Emails (B) Infected websites (C) **Official Antivirus CDs** (D) USBs
5. In which year Apple II virus came into existence?

    (A) Apple 1 (B) Linux (C) Microsoft (D) **Apple II**.

Answers

1. 10
2. Trojans
3. Virus
4. Official Antivirus
5. Apple II.

Summary Question

1. List the file analyzer malware tool for validation?
2. What do you mean by regshot and process explorer malware analyzing tool?
3. Can you list the advantages of antivirus in malware detection techniques?
4. List the importance of firewall and its importance based on analyzing tool?
5. Detail about security information and event management?

Chapter 7
Firewalls

A firewall is nothing but the firmware or software used to restrict unauthorized access to the network. The firewall has rules to inspect the incoming traffic and outgoing traffics and block unwanted threats. Therefore, the firewall is the main component of network security, which comes from the build-in of several systems such as Linux, Windows, and Mac. The discussed firewall has been utilized in both enterprise and personal settings.

7.1 Importance of Firewalls

The firewall placed an essential role in network security; a few firewall importance are listed.

- Securing computers from unwanted or unauthorized access.
- Analyzing and blocking the unwanted contents.
- Prevent content from various malware, viruses, and worms.
- Securing network in the multi-person environment.
- It helps to secure private and confidential information.
- Creating a protected network in multi-user interactions such as online video gaming.
- Alerting the illegal outbound and illegal inbound attempts.
- Block unsolicited content.
- Preventing or blocks the hacking attempts from cybercriminals.

© The Author(s), under exclusive license to Springer Nature Singapore Pte Ltd. 2023   117
B. S. Rawal et al., *Cybersecurity and Identity Access Management*,
https://doi.org/10.1007/978-981-19-2658-7_7

7.2 Uses of Firewalls

Firewalls are utilized in both consumer and corporate settings. Modern organizations are integrated with security information and event management (SIEM) processes with cybersecurity devices. They may be put in at associate organization's network perimeter to protect against external threats or among the network to make segmentation and guard against corporate executive threats.

In addition to immediate threat defense, firewalls perform necessary work and audit functions. They keep a record of events, which directors might utilize to spot patterns and improve rule sets. Rules ought to be updated often to stay up with ever-evolving cybersecurity threats. Vendors discover new threats and develop patches to hide them as before long as attainable.

In a single home network, a firewall will filter traffic and alert the user to intrusions. They're incredibly helpful for always-on connections, like digital telephone lines (DSL) or electronic cable equipment, resulting from those affiliation sorts use static information science addresses. They're usually used aboard to antivirus applications. Personal firewalls, not like company ones, area units typically one product as against a group of varied merchandise. They will be a package or a tool with a firewall computer code embedded. Hardware/firmware firewalls area unit is usually used for setting restrictions between in-home devices.

7.3 How Does Firewall Works

Firewalls create the border between the network and the external network. The created border helps to examine the leaving and entering networks. During this process, pre-defined rules are used to explore standard and malicious packets.

In addition to immediate threat defense, firewalls perform vital work and audit functions. They keep a record of events, which directors might utilize to spot patterns and improve rule sets. Rules ought to be updated often to stay up with ever-evolving cybersecurity threats. Vendors discover new threats and develop patches to hide them as presently as attainable.

In a single home network, a firewall will filter traffic and alert the user to intrusions. They're incredibly helpful for always-on connections, like digital connective (DSL) or electronic cable equipment, resulting from those association sorts use static science addresses. They're typically used aboard to antivirus applications. Personal firewalls, not like company ones, a square measure sometimes one product as critical a group of the assorted work. They'll be a software package or a tool with a firewall microcode embedded. Hardware/firmware firewalls square measure typically used for setting restrictions between in-home devices.

The term "packets" refers to items of knowledge that area unit formatted for web transfer. Moreover, packets contain the information itself as info regarding the data, like wherever it came from. Firewalls will use this packet info to determine whether

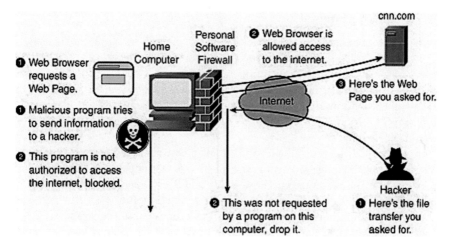

Fig. 7.1 Working processes of firewalls

or not a given packet abides by the ruleset. If it doesn't, the package is going to be barred from getting into the guarded network. Rule sets are often supported by many things indicated by packet information, including source, destination, and content.

These characteristics are also depicted otherwise at entirely different levels of the network. As a packet travels through the system, it's reformatted many times to inform the protocol wherever to send it. Differing types of firewalls exist to browse packages at completely different network levels. Then the firewall working process is illustrated in Fig. 7.1.

7.4 Types of Firewall

Firewalls area unit either classified by the means they filter information or by the system they shield. When categorizing by what they shield, the two sorts are network-based and host-based. Network-based firewalls guard entire networks and area units, usually hardware. Host-based firewalls guard individual devices—called hosts—and area units usually package. When categorizing by filtering methodology, the most sorts are:

- Packet filtering firewall
- State full infection firewall
- Proxy firewall
- Next-generation firewall.

Packet Filtering Firewall

When a packet permits through a packet-filtering firewall, its supply and destination address, protocol, and destination port range square measure checked. The packet is born—that means not forwarded to its destination—if it doesn't accommodate the firewall's rule set. For instance, if a firewall is designed with a rule to dam Telnet access, then the firewall can drop packets destined for Transmission Management Protocol (TCP) port range twenty-three, the port wherever a Telnet server application would be listening.

A packet-filtering firewall works in the main network layer of the OSI reference model, though the transport layer is employed to get the supply and destination port numbers. It examines every packet severally and doesn't apprehend whether or not any given packet is an element of an existing stream of traffic.

The packet-filtering firewall is effective, however as a result of it processes every packet in isolation, it may be susceptible to IP spoofing attacks and has mostly been replaced by stateful scrutiny firewalls. According to the discussion, the packet filtering firewall is illustrated in Fig. 7.2.

Stateful Inspection Firewall

Stateful review firewalls—additionally referred to as dynamic packet-filtering fire-walls—monitor communication packets over time and examined each arriving and outward-bound package.

This type preserves a table that keeps track of all open associates. Once new packets arrive, it compares data within the packet header to the state table—its list of valid associations—and determines whether or not the packet is a component of a longtime association. If it is, the package is let through while not additional analysis. If the packet doesn't match the existing associate association, it's evaluated in step with the rule set for brand new connections.

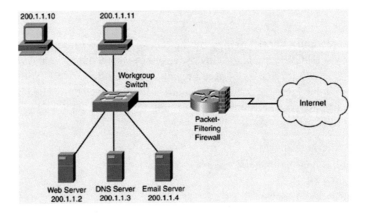

Fig. 7.2 Packet filtering firewalls

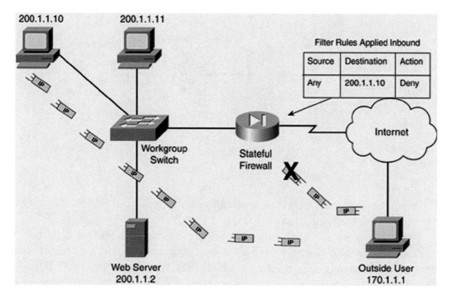

Fig. 7.3 Stateful inspection firewall

Although stateful review firewalls area units are quite effective, they'll be at risk of denial-of-service (DoS) attacks. DoS attacks work by taking benefit of proven connections that this kind typically assumes area unit safe. Then the stateful inspection firewall is illustrated in Fig. 7.3.

Proxy Firewall

This type may additionally be spoken as a proxy-based or reverse-proxy firewall. They supply application layer filtering and might examine the payload of a packet to tell apart valid requests from malicious code disguised as a sound request for information. As attacks against net servers became additional common, it became apparent that there was a requirement for firewalls to guard networks from attacks at the appliance layer. Packet-filtering and stateful review firewalls cannot do that at the appliance layer.

Since this kind examines the payload's content, it provides security engineers additional granular management over network traffic. As an example, it will permit or deny a particular incoming Telnet command from a specific user, whereas different varieties will solely management general incoming requests from a specific host.

When this kind lives on a proxy server—creating it a proxy firewall—it makes it tougher for Associate in Nursing assaulter to get wherever the network really is and creates one more layer of security. Each the consumer and also the server square measure forced to conduct the session through Associate in Nursing negotiate—the proxy server that hosts Associate in Nursing application layer firewall. Whenever an Associate in Nursing external consumer requests an affiliation to an inside server or the other way around, the consumer can open a reference to the proxy instead.

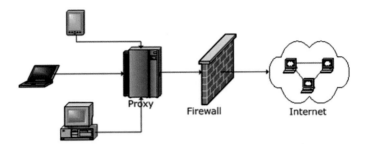

Fig. 7.4 Proxy firewall

If the affiliation request meets the standards within the firewall rule base, the proxy firewall can open affiliation to the requested server.

The key advantage of application-layer filtering is that the ability to dam specific content, like notable malware or bound websites, and acknowledge once bound applications and protocols, like machine-readable text Transfer Protocol (HTTP), File Transfer Protocol (FTP), and name system (DNS), the square measure being used. Application layer firewall rules may be wont to manage the execution of files or the handling of information by specific applications. Then the proxy firewall structure is illustrated in Fig. 7.4.

Next-Generation Firewall (NGFW)

This type could be a combination of the opposite sorts with extra security software packages and devices bundled in. Every kind has its own strengths and weaknesses, some shield networks at totally different layers of the OSI model. The advantage of an NGFW is that it combines the strengths of every kind of cowl, each type's weakness. AN NGFW is commonly a bundle of technologies below one name as critical one element.

Modern network perimeters have such a lot of entry points and differing kinds of users that stronger access management and security at the host area unit are needed. This want for a multilayer approach has diode to the emergence of NGFWs. An NGFW integrates three key assets: ancient firewall capabilities, application awareness, ANd an IPS. Just like the introduction of stateful review to first-generation firewalls, NGFWs bring extra context to the firewall's decision-making method Table 7.1.

NGFWs mix the capabilities of ancient enterprise firewalls—together with Network Address Translation (NAT), Uniform Resource surveyor (URL) obstruction, and virtual non-public networks (VPNs)—with quality of service (QoS) practicality and options not historically found in first-generation merchandise. NGFWs support intent-based networking by together with Secure Sockets Layer (SSL) and Secure Shell (SSH) review and reputation-based malware detection. NGFWs conjointly use deep packet review (DPI) to visualize the contents of packets and forestall malware. When an NGFW or any firewall is employed in conjunction with different

Table 7.1 Types of firewall

Types	Descriptions
Proxy firewall	An early style of firewall device, a proxy firewall, is the entree from one network to a different for a selected application. Proxy servers will give extra practicality like content caching and security by preventing direct connections from outside the network. However, this additionally might impact output capabilities and also the applications they'll support
Stateful inspection	Now thought of as a "traditional" firewall, a stateful review firewall permits or blocks traffic supported state, port, and protocol. It monitors all activity from the gap of an association till it's closed. Filtering selections are created to support each administrator-defined rule also as context, which refers to mistreatment info from previous associations and packets happiness to constant connection
Unified threat management firewall	A UTM device usually combines, in an exceedingly loosely coupled approach, the functions of a stateful examination firewall with intrusion bar and antivirus. It is going to additionally embody further services and infrequently cloud management. UTMs target simplicity and simple use
Next generation firewalls (NGFW)	Firewalls have evolved on the far side, with easy packet filtering and stateful scrutiny. Most corporations' area units were deploying next-generation firewalls to fashionable dam threats like advanced malware and application-layer attacks. According to Gartner, Inc.'s definition, a next-generation firewall should include standard firewall capabilities like stateful scrutiny, integrated intrusion interference, application awareness, and management to visualize and block risky apps, upgrade methods to incorporate future info feeds, Techniques to handle evolving security threats. While these capabilities area unit is progressively changing into the quality for many corporations, NGFWs will do a lot of
Threat-focused NGFW	These firewalls embody all the capabilities of a conventional NGFW and conjointly give advanced threat detection and correction. With a threat focused NGFW, you can Know that assets area unit most in danger with complete context awareness Quickly react to attacks with intelligent security automation that sets policies and hardens your defenses dynamically Better sight evasive or suspicious activity with network and end event correlation. Greatly decrease the time from detection to cleanup with retrospective security that incessantly monitors for suspicious activity and behavior even once the initial examination Ease administration and scale back quality with unified policies that shield across the whole attack time

(continued)

Table 7.1 (continued)

Types	Descriptions
Virtual firewall	A virtual firewall is usually deployed as a virtual appliance during a personal cloud (VMware ESXi, Microsoft Hyper-V, KVM) or public cloud (AWS, Azure, Google) to watch and secure traffic across physical and virtual networks. A virtual firewall is commonly a key element in software-defined networks (SDN)

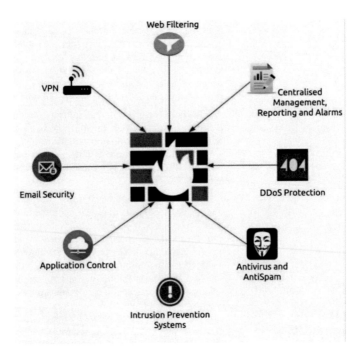

Fig. 7.5 Next-generation firewall

devices, it's termed unified threat management (UTM). According to the discussion, next-generation firewall is shown in Fig. 7.5.

7.5 Benefits of Firewalls

The right understanding of firewall benefits helps to improve the business process because it depends on several networks, operations, and technologies. The main goal of the firewall is to avoid and eliminate external threats, hackers, and malware activities to access the information in systems. A firewall has the following benefits:

- Helps to monitor network traffics
- Eliminates and stop virus attacks
- Prevent hacking activities
- Stops spyware
- Increase privacy and security.

7.6 Advantages and Disadvantages of Firewalls

Firewalls are having several advantages, which are listed to understand why the firewall techniques are utilized to control the hacker, malware, and virus-related activities.

- Versatility
- Intelligent port control
- Simple infrastructure
- Updated threat protection
- Consistent network speed.

As same as advantages, a firewall has few disadvantages that are listed to get the improvements in the firewall-based malware, hacker, and virus detection process

- Application awareness limitations
- Issues with network speed
- Logistic drawback
- Lack of evaluation capabilities.

7.7 Firewall Threats and Vulnerability

In cybersecurity architecture, firewall placed an important role, but firewall only does not provide complete solutions. The firewall also has a few threats and vulnerabilities, which are listed as follows.

- Insider Attacker
- Missed security patches
- Configuration mistakes
- Lack of deep packet inspection
- Distributed Denial-of-service attacks (DDoS).

Insider Attacker

An edge firewall is intended to ward off assaults that begin from outside of your organization. Anyway, what happens when the assault begins from within? Regularly, the border firewall gets futile—overall, the aggressor is as of now on your framework. Nonetheless, in any event, when an assault begins from inside your organization, firewalls can benefit a few—IF you have inner firewalls on the head of your edge firewalls. Interior firewalls help to segment singular resources on your organization, so assailants need to work more earnestly to move to start with one framework then onto the next one. This helps increment the aggressor's breakout time, so you have more opportunity to react to the assault.

Missed Security Patches

This is an issue that emerges when network firewall programming isn't overseen appropriately. For any product program, there are weaknesses that aggressors may misuse—this is as valid for firewall programs for what it's worth of some other bit of programming. At the point when firewall sellers find these weaknesses, they, as a rule, work to make a fix that fixes the issue at the earliest opportunity. Notwithstanding, the fix's simple presence doesn't imply that it will consequently be applied to your organization's firewall program. Until that fix is really applied to your firewall programming, the weakness is still there—simply holding on to be abused by an irregular assailant. The best fix for this issue is to make and adhere to a severe fix the executive's plan. Under such a timetable, you (or the individual dealing with your network safety) should check for any security refreshes for your firewall programming and try to apply them as quickly as time permits.

Configuration Mistakes

In any event, when a firewall is set up on your organization and has the entirety of the most recent weakness patches, it can, in any case, cause issues if the firewall's design settings make clashes. This can prompt lost execution on your organization at times and a firewall altogether neglecting to give assurance in others. For instance, dynamic steering is a setting that was some time in the past considered a poorly conceived notion to empower in light of the fact that it brings about lost control that decreases security. However, a few organizations leave it on, making a weakness in their firewall security. Having an inadequately designed firewall is somewhat similar to filling a stronghold's canal with sand and placing the way into the fundamental door in stow away a-key right close to the passage—you're simply making things simpler for assailants while sitting around idly, cash, and exertion on your "security" measure.

Lack of Deep Packet Inspection

Layer 7 assessment is a thorough review mode utilized by cutting edge firewalls to look at the substance of a data parcel before endorsing or rejecting that parcel section to or from a framework. Less progressed firewalls may essentially check the information bundle's place of inception and objective before affirming or denying a solicitation—data that an aggressor can without much of a stretch parody to deceive your organization's firewall. The best fix for this issue is to utilize a firewall that can perform profound parcel reviews to check data bundles for known malware, so it tends to be dismissed.

Distributed Daniel of Services

DDoS assaults are a regularly utilized assault procedure noted for being exceptionally compelling and generally minimal effort to execute. The essential objective is to overpower a safeguard's assets and cause a closure or delayed failure to convey administrations. One classification of assault—convention assaults—is intended to deplete firewall and burden balancer assets to shield them from preparing authentic traffic. While firewalls can moderate a few sorts of DDoS assaults, they can even now be over-burden by convention assaults.

There is no simple fix for DDoS assaults, as there are various assault methodologies that can use various shortcomings in your organization's design. Some online protection specialist organizations offer "cleaning" administrations, wherein they redirect approaching traffic away from your organization and sort out the authentic access endeavors from the DDoS traffic. This authentic traffic is then shipped off your organization so you can continue ordinary tasks.

Alone, firewalls can't shield your organization from the entirety of the dangers that are out there. In any case, they can fill in as a vital aspect of a bigger network safety system to shield your business. Need to study how you can create a solid network protection plan for your business? Look at our free manual for network protection fundamentals at the connection beneath! Then again, contact Comp quip Cybersecurity currently to get master exhortation from an accomplished network safety proficient.

Summary

This section explains the detailed view of firewall, importance, benefits, working process, advantages, and disadvantages are discussed. The general explanation helps to understand that how the firewall system helps to maintain security in the network system. In addition to these threats and vulnerabilities are discussed to understand the limitation of firewalls.

Assignment Question

1. List the importance of firewall
2. What are all the uses of firewall
3. List the types of Firewall
4. Can you illustrate the advantages and disadvantages of firewalls
5. What do you mean by threats?

Multiple Choices

1. .... Is the direction virus.

    (A) Multipartite virus (B) Boot sector virus (C) resident virus (D) **Non-resident virus**

2. .... Affects the .exe files and boots strap.

    (A) **Multipartite virus** (B) Boot sector virus (C) Poly virus (D) Non-resident virus

3. …… affects and deletes all the files in the system.

    (A) **Overwrite virus** (B) Resident sector virus (C) Poly virus (D) Non-resident
    virus

4. …… is called cavity virus.

    (A) Multipartite virus (B) Boot sector virus (C) **Space filler virus** (D) Non-
    resident virus

5. List the number of viruses that hide in different ways……

    (A) 1 (B) 2 (C) **3** (D) 4

Answers

(A)   Non-resident virus
(B)   Multipartite virus
(C)   Overwrite virus
(D)   Space Filler Virus
(E)   3.

Summary Question

1. What are the current firewalls that prevail at the present web level?
2. Illustrate the benefits of a firewall?
3. Why are companies more significant in using a firewall?
4. What are all the necessary factors which are needed to make the firewall
    effective?
5. What are all the advantages and disadvantages of security software on the web?

Chapter 8
Cryptography

Cryptography is a technique of securing info and communications through the use of codes so solely those people for whom the data is meant will are aware of it and method it. Therefore, preventing unauthorized access to info. The prefix "crypt" suggests that "hidden," and the suffix graph suggests that "writing."

In cryptography, the techniques that area unit used to safeguard info area unit obtained from mathematical ideas and a collection of rules primarily based calculations referred to as algorithms to convert messages in ways in which create it onerous to decipher it. These algorithms area unit used for scientific discipline key generation, digital sign language, verification to safeguard information privacy, internet browsing on the web, and to safeguard confidential transactions like MasterCard and open-end credit transactions.

8.1 Evaluation of Cryptography

The art of cryptography is taken into account to turn at the side of the art of writing. As civilizations evolved, groups of people got organized into tribes, groups, and kingdoms. This junction rectifier to the emergence of concepts like power, battles, supremacy, and politics. These concepts any oxyacetylene the natural want of individuals to speak on the Q.T. with the selective recipient that successively ensured the continual evolution of cryptography in addition. The roots of cryptography are found in Roman and Egyptian civilizations. During and when the Renaissance, varied Italian and apostolical states junction rectified the speedy proliferation of cryptanalytic techniques. Varied analysis and attack techniques were researched during this era to interrupt the key codes.

Improved committal to writing techniques like Vigenere committal to writing came into existence within the fifteenth century, that offered moving letters within the message with a variety of variable places rather than moving them an equivalent number of places. Only when the nineteenth century, cryptography evolved from the

© The Author(s), under exclusive license to Springer Nature Singapore Pte Ltd. 2023
B. S. Rawal et al., *Cybersecurity and Identity Access Management*,
https://doi.org/10.1007/978-981-19-2658-7_8

impromptu approaches to secret writing to a lot of refined art and science of data security.

In the early twentieth century, the invention of mechanical and mechanical device machines, like the Enigma rotor machine, provided a lot of advanced and economic suggestions that of committal to writing the knowledge. During the amount of warfare II, each cryptography and science became too mathematical. With the advances going down during this field, government organizations, military units, and a few company homes started adopting the applications of cryptography. They used cryptography to protect their secrets from others. Now, the arrival of computers and, therefore, the net has brought effective cryptography inside the reach of folk.

8.2 Features of Cryptography

Cryptography has several features that are used to ensure the security of the data. The detailed cryptography features are shown in Fig. 8.1.

- Confidentiality:
  - Information will solely be accessed by the person for whom it's meant, and no different person except him will access it.
- Integrity:
  - Information can't be changed in storage or transition between sender and meant receiver with no addition to info being detected.
- Non-repudiation:
  - The creator/sender cannot deny his or her intention to send information at a later stage.
- Authentication:
  - The identities of the sender and receiver area unit were confirmed. Yet as the destination/origin of data is confirmed.

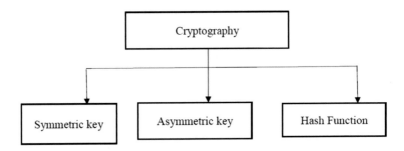

Fig. 8.1 Types of cryptography

8.3 Types of Cryptography

Cryptography is divided into three types that are shown in Fig. 8.1.

Symmetric Key Cryptography

It is an associate cryptography system wherever the sender and receiver of a message use one common key to write and decipher messages. Bilaterally symmetric key systems area unit quicker and less complicated however the matter is that sender and receiver need to somehow exchange key in a very secure manner. The foremost widespread bilaterally symmetric key cryptography system is the encoding system (DES) Fig. 8.2.

Hash Functions

There is no usage of any key during this rule. A hash price with fastened length is calculated as per the plain text that makes it not possible for contents of plain text to be recovered. Several operative systems use hash functions to write passwords Fig. 8.3.

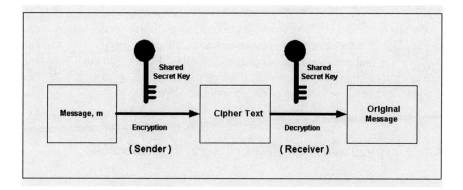

Fig. 8.2 Symmetric key cryptography

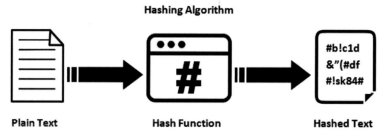

Fig. 8.3 Hash function-based cryptography

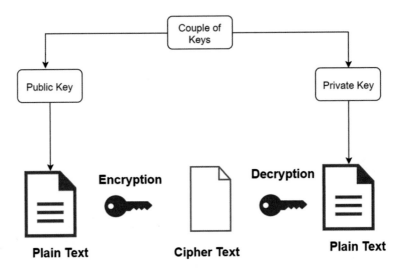

Fig. 8.4 Asymmetric cryptography

Asymmetric Key Cryptography

Under this method, a combination of keys is employed to write and decipher info. A public secret is used for cryptography, and a personal secret is used for secret writing. A public key and personal Key area unit completely different. Although the general public secret's familiar by everybody, the meant receiver will solely decipher it as a result of, and he alone is aware of the personal key Fig. 8.4.

8.4 Cryptography and Network Security Principles

In a gift day situation, the security of the system is that the sole priority of any organization. The biggest aim of any organization is to guard its knowledge against attackers. In cryptography, attacks are of two sorts like passive attacks and active attacks.

Passive attacks are people who retrieve data from the system while not poignant the system resources, whereas active attacks are people who retrieve system data and build changes to the system resources and their operations. The principles of security are classified as follows.

Confidentiality

The degree of confidentiality determines the secrecy of the knowledge. The principle specifies that solely the sender and receiver are able to access the knowledge shared between them. Confidentiality compromises if an associate nursing unauthorized person is ready to access a message.

For example, allow us to think about sender A needs to share some with receiver B, and also the information gets intercepted by the wrongdoer C. Currently, the wind is within the hands of Associate in Nursing trespasser C.

Authentication

Authentication is that the mechanism to spot the user or system, or entity. It ensures the identity of the person attempting to access the knowledge. The authentication is usually secured by victimization username and positive identification. The licensed person whose identity is preregistered will prove his/her identity and may access the sensitive data.

Integrity

Integrity offers the reassurance that the knowledge received is actual and correct. If the content of the message is modified once the sender sends it, however, before reaching the supposed receiver, then it's aforementioned that the integrity of the message is lost.

Non-repudiation

Non-repudiation could be a mechanism that stops the denial of the message content sent through a network. In some cases, the sender sends the message and later denies it. However, the non-repudiation doesn't enable the sender to refuse the receiver.

Access Control

The principle of access management is set by role management and rule management. Role management determines World Health Organization ought to access the info, whereas rule management determines up to what extent one will access the info. The knowledge displayed depends on the one who is accessing it.

Availability

The principle of accessibility states that the resources are out there to authorize parties in the least time. Data will not be helpful if it's not out there to be accessed. Systems ought to have adequate accessibility of data to satisfy the user's request.

8.5 Cryptographic Algorithms

In this IoT domain, security matters foremost. Although there are several security mechanisms in observation, they are doing not hold the power to come back up with current day good applications primarily for the software system operational with resource-constraint instrumentation. In an exceeding consequence of this, cryptography algorithms came into observe guaranteeing increased security. So, a few of the cryptographic algorithms are as follows:

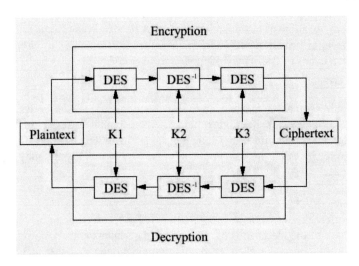

Fig. 8.5 Tripe DES cryptography algorithm

Triple DES

Taking over from the traditional DES mechanism, triple DES was presently enforced within the security approaches. These algorithms allow hackers to ultimately gained the information to beat in a straightforward approach. This was an extensively enforced approach by several of the enterprises. Triple DES operates with three keys having fifty-six bits per key. The complete key length could be a most of bits, whereas consultants would contend that 112-bits in key intensity is a lot of probable. This algorithmic program handles to form a reliable hardware secret writing declare banking facilities and additionally for different industries Fig. 8.5.

Blowfish

To replace the approaches of Triple DES, Blowfish was principally developed. This secret writing algorithmic rule goes different ways messages into clocks having sixty-four bits and encrypts these clocks one by one. The charming feature that lies in Blowfish is its speed and efficaciousness. As this can be Associate in Nursing open algorithmic rule for everybody, several gained the advantages of implementing this. Each scope of the IT domain, starting from software package to e-commerce, is creating use of this algorithmic rule because it shows in depth options for countersign protection. Of these enable this algorithmic rule to be most distinguished within the market.

RSA

One of the public-key coding algorithms accustomed to encode data transmitted through the web. It had been a widely used rule in GPG and PGP methodologies. RSA is classed underneath bilateral kinds of algorithms because it performs its operation

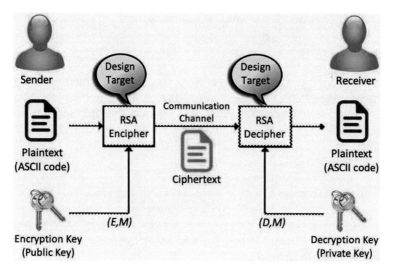

Fig. 8.6 RSA algorithm

employing a few keys. One among the keys is employed for coding and also the alternative for decoding functions Fig. 8.6.

Twofish

This rule implements keys to produce security, and because it comes underneath the bilateral methodology, only one secret's necessary. The keys of this rule square measure with the utmost length of 256 bits. Of the foremost offered algorithms, Twofish is especially well-known for its speed and ideal to be enforced within the hardware and software package applications. Also, its associate degree overtly accessible rule and has been in execution by several.

AES (Advanced Encryption Standard)

This is the foremost trusty rule technique by the U.S. administration and plenty of alternative enterprises. However, this works expeditiously in 128-bit coding type, 192- and 256-bits square measure chiefly used for Brobdingnagian coding activities. Being thus defendable to any or all hacking systems, the AES technique receives in depth commendation for encrypting data within the non-public domain.

8.6 Tools for Cryptography

Cryptography tools square measure additional helpful within the things of signature confirmation, code linguistic communication, and to perform alternative cryptography activities. Here square measures the extensively used cryptography tools.

Security Token

This token is used to verify the user. A security token is meant to be encrypted to perform a protected exchange of data. Also, it provides complete tastefulness for the communications protocol. So, the server-side developed token is used by a browser to travel on with the state. In general, it's the tactic that moves with remote authentication.

JCA

This is the tool that wants to authorize the coding method. This tool could be termed a Java cryptanalytic libraries. These Java libraries square measure enclosed with predefined activities wherever those got to be foreign before implementation. Though it's the Java library, it works in proportion with alternative frameworks and so supports the event of multiple applications.

SignTool .exe

This is the favored tool principally employed by Microsoft to sign the files. Adding a signature and time stamp to any reasonably file is that the distinguishing feature supported by this tool. With the timestamp within the file, it holds power to certify the file. The entire feature in SignTool .exe ensures increased reliability of the file.

Docker

Using stevedore, one will build immense applications. The data maintained within the stevedore is totally in associate degree encrypted format. In this, cryptography should be strictly followed to maneuver with the coding of knowledge. What is more, each file and data square measure encrypted so permitting nobody to access the items having no precise access key. Stevedore is additionally contemplated as cloud storage permitting users to manage the data either on an infatuated or shared server.

CertMgr .exe

This is the installation file because it is in .exe-extension format. CertMgr holds sensible for the management of varied certificates. Together with this, it even handles CRLs wherever those square measure certificate revocation lists. The target of cryptography in certificate development is to make sure that the data that's changed between the parties is additional protected, and this tool supports to feature of extra bits in protection.

8.7   Advantages and Disadvantages of Cryptography

Cryptography is a necessary info security tool. It provides the four most elementary services of data.

- Confidentiality—Secret writing technique will guard the knowledge and communication from unauthorized revelation and access of data.
- Authentication—The cryptanalytic techniques like mackintosh and digital signatures will defend info against spoofing and forgeries.
- Data Integrity—The cryptanalytic hash functions square measure enjoying a very important role in reassuring the users regarding the information integrity.
- Non-repudiation—The digital signature provides the non-repudiation service to protect against the dispute which will arise because of denial of passing message by the sender.

All these elementary services offered by cryptography have enabled the conduct of business over the networks victimization of the PC systems in an extraordinarily economical and effective manner.

Disadvantage of Cryptography

Apart from the four elementary components of knowledge security, there are a unit alternative problem that has an effect on the effective use of knowledge:

- A powerfully encrypted, authentic, and digitally signed data is troublesome to access even for a legitimate user at an important time of decision-making. The network or the PC system is attacked and rendered non-functional by an unwelcome person.
- High accessibility, every one of the basic aspects of knowledge security, cannot be ensured through the utilization of cryptography. Alternative ways area unit required to protect against the threats like denial of service or complete breakdown of knowledge system.
- Another elementary would like of knowledge security of selective access management conjointly cannot be accomplished through the utilization of cryptography. Body controls and procedures area unit needed to be exercised for a similar.
- Cryptography doesn't guard against the vulnerabilities and threats that emerge from the poor style of systems, protocols, and procedures. These ought to be fastened through the correct style and putting in of a defensive infrastructure.
- Cryptography comes at a value. The price is in terms of your time and cash. The addition of cryptographical techniques within the information science ends up in delay. The use of public key cryptography needs putting in and maintenance of public key infrastructure requiring a handsome monetary budget.
- The security of graphical crypto technique is predicated on the process issue of mathematical issues. Any breakthrough in determining such mathematical issues or increasing the computing power will render a cryptographic technique vulnerable.

8.8 Applications of Cryptography

Conventionally, cryptography was in implementation just for securing functions. Wax seals, hand signatures, and few different kinds of security ways were typically used to create certain dependableness and accuracy of the transmitter. And with the arrival of digital transmissions, security becomes a lot essential than cryptography mechanisms began to outstrip its utilization for maintaining utmost secrecy. Many of the applications of cryptography are mentioned below.

To Maintain Secrecy in Storage

Cryptography permits storing the encrypted information allowing users to remain back from the main hole of escape by hackers.

Reliability in Transmission

A conventional approach that permits dependableness is to hold out a verification of the communicated info then communicate the corresponding verification in an associate encrypted format. Once each verification and encrypted information is received, the information is once more verified and compared to the communicated checksum when the method of decipherment. Thus, effective cryptologic mechanisms are a lot of crucial to assure dependableness in message transmission.

Authentication of Identity

Cryptography is powerfully coupled to the approach of mistreatment passwords, and innovative systems most likely create use of sturdy cryptologic ways alongside the physical ways of people and collective secrets, providing extremely reliable verification of identity.

Summary

Thus, the chapter is discussing the basic information about cryptography, features, principles to maintaining data security. In addition to this, different cryptographic algorithms, tools, applications, advantages, and disadvantages are discussed to improve information security.

Assignment Question

1. How do you evaluate cryptography?
2. What are all the techniques available in cryptography?
3. List the approaches used in the cryptographic analysis?
4. What are all the tools which used in cryptographic techniques?
5. What are all the advantages and disadvantages of cryptographic techniques?

Multiple Choice Questions

1. ....Algorithm used in asymmetric key cryptography.

    (A) RSA (B) Diffie Hellman (C) Electronic Code (D) DSA

2.   What do you mean by data encryption standard….

     (A) Block cipher (B) Stream cipher (C) Bit cipher (D) Byte cipher
3.   In cryptography, the shuffling of letters is named as………..

     (A) trans-positional approach (B) substitution approach (C) both A and B (D) Quadratic ciphers
4.   Cryptanalysis is used ……….

     (A) Find insecurity on cryptographic scheme (B) Improve speed (C) For encryption (D) To create new ciphers
5.   ElGamal encryption system is ……….

     (A) Asymmetric algorithm (B) symmetric algorithm (C) Encryption algorithm (D) Cipher method

Answers

1.   RSA
2.   Block Cipher
3.   Trans-positional approach
4.   Find insecurity on cryptographic scheme
5.   Symmetric algorithm.

Summary Question

1.   Mention the importance of cryptographic algorithms and its application
2.   List the network security principles and their application
3.   Classification models for the features of cryptography need to mention based on security principles
4.   Mention the real time application of cryptography and its importance
5.   Mention the limitation of cryptographic principles and their application.

Chapter 9
Control Physical and Logical Access to Assets

9.1 Managing Access to Assets

Managing access to assets is the prime theme of security, and access control is provided by many different security controls working together. Mostly, an asset conforms of information, systems, devices, facilities, and personnel which need to be secured.

Information

Data contains the organization's information, which can be stored in simple files on computers, servers, and smaller devices. Information can also be stored on a large database system within a server farm. To prevent unauthorized access to information, access control is required here.

Systems

Information technology (IT) systems of an organization include providing one or more services that need to be secured to prevent unwanted intrusion. Like user files are stored in a simple file server, which is a system. Also, e-commerce service providing a database server working with a web server is a system.

Devices

Any computing system, including smartphones, desktop computers, servers, portable laptop computers, tablets, and external devices such as printers, all are called devices. Bring your device (BYOD) policies are being adopted by more and more organizations allowing employees to connect their personally owned device to an organization's network. Organizational data stored on the devices is still needed to be secured as an asset of the organization, though the devices are the property of their owners.

© The Author(s), under exclusive license to Springer Nature Singapore Pte Ltd. 2023   141
B. S. Rawal et al., *Cybersecurity and Identity Access Management*,
https://doi.org/10.1007/978-981-19-2658-7_9

Facilities

The physical location of an organization conforms to its facilities that include any place that it owns or rents. This includes individual rooms, entire buildings, or entire complexes of several buildings. Physical security is required to protect facilities.

Personnel

Employees of an organization are also valuable assets to an organization. Primary ways of securing them in the workplace are to ensure adequate safety measures to protect them from getting injured or died of accidents.

Subjects Versus Objects

Access control not only just controls users' access to files or services but also many more. It is related to entities like subjects and objects. The transfer of information from a subject to an object is known as access, which defines both subjects and objects more precisely.

Subject

An active entity, subject accesses a passive object to receiving data from, or information about, an object. Subjects include users, computers, programs, processes, or anything else that can access a resource. After authorization, subjects can modify objects.

Object

An active subject gets information from a passive entity called an object. Examples of objects include programs, processes, files, databases, computers, printers, and storage media.

9.2 Why Access Control is Required: The CIA Triad

The basic reason why organizations implement access control mechanisms is to prevent losses of information and resources. If you categorize organizations' loss, it can be divided into three categories: loss of confidentiality, loss of availability, and loss of integrity. These losses are so fundamental to IT security that they are frequently referred to as the CIA triad or sometimes the security triad or AIC triad.

Confidentiality

Loss of confidentiality results when unauthorized entities access into system or data. Access controls help to pass only authorized subjects to access objects to retain confidentiality.

Integrity

Integrity ensures that modification of data or system configurations is not done without authorization. If there is an unauthorized change, security controls detect the change immediately to sustain confidentiality. If, unfortunately, unwanted or unauthorized changes occur, it causes a loss of integrity.

Availability

Authorized subjects must be granted to search objects within a reasonable amount of time. It can be said that data and systems should be available in need of users and other subjects. If the systems failed to provide viable information on time, then it will be regarded as non-operational, or the data is inaccessible, resulting in unavailability.

9.3 Classification of Access Control

Normally, access control is any software, hardware, or administrative procedures or policies that control access to systems, devices, and resources. The goal of the effectuation of access control is to provide access to authorized subjects to prevent troublemaking unauthorized access attempts.

The following overall steps include access control:

1. Determination of whether the access is authorized.
2. Identification and authentication of authorized users or subjects intending to access systems and resources.
3. Restrict or grant access based on the subject's identity authorization.
4. Trace, follow, and record access attempts.

Varieties of controls are involved in these steps. The three basic control types are a preventive, detective, and corrective. Security control means you want to prevent any type of possible security breach or incident. But it's the harsh truth that unwanted events occur despite security controls, countermeasures, and safeguards. When this happens, you want to detect the event as soon as possible to prevent more loss. After detection, you want to correct it. In that case, there are also four other access control types to handle this type of situation as a deterrent, recovery, directive, and compensation access controls.

But if you want to have maximum benefit from security mechanisms, security controls, countermeasures, and safeguards must be implemented in a defense-in-depth manner administratively, logically/technically, or physically.

9.3.1 Preventive Access Control

Preventive control attempts to dismiss or stop unauthorized or unwanted security breaches from occurring.

Examples of preventive access controls include mantraps, fences, biometrics, locks, alarm systems, lighting, job rotation policies, separation of duties policies, penetration testing, data classification, encryption, access control methods, the presence of security cameras or closed-circuit television (CCTV), smartcards, auditing, security policies, callback procedures, antivirus software, firewalls, security awareness training, and intrusion prevention systems.

9.3.2 Detective Access Control

Detective control involves discovering or detecting unauthorized or unwanted activity before breaching. It can operate and discover the activity only after the anomaly has occurred.

Examples of detective access controls cover recording and reviewing of events captured by CCTV, security guards, supervision and reviews of users, motion detectors, intrusion detection systems, mandatory vacation policies, job rotation policies, audit trails, honeypots, or honeynets, violation reports, and incident investigations.

9.3.3 Corrective Access Control

A corrective control redesigns or modifies the environment to return systems to normal after an unwanted or unauthorized security breach has occurred. Corrective controls simply attempt to correct any incident that occurred as a result of a security breach.

Examples of corrective controls include rebooting a system or terminating malicious activity, antivirus solutions to remove or quarantine virus in a system, active unauthorized mobility detection systems that can refurbish the environment to stop an attack in progress, backup and restoration plan to ensure that if an unfortunate breach happens then lost data can be restored.

9.3.4 Deterrent Access Control

Deterrent control involves discouraging security policy violations. Preventive and Deterrent controls are quite similar, except deterrent controls mostly depend on individuals' decisions on causing an unwanted action. On the contrary, preventive control dismisses the action before being happened.

Examples of deterrent control include locks, keycards, fences, policies, security awareness training, security cameras, security badges, mantraps, and guards.

9.3.5 Recovery Access Control

Recovery access control intends to restore or repair damaged resources, systems, and capabilities after a security breach. Recovery controls have more advanced or complex abilities than corrective controls, which is although an extension of that.

Examples of recovery access controls cover fault-tolerant drive systems, backups and restores, system imaging, antivirus software, server clustering, and virtual machine or database shadowing.

9.3.6 Directive Access Control

A directive control intends to control, direct, or confine the actions of subjects to follow or encourage compliance with security features and policies.

Examples of directive access controls cover supervision, monitor, security policy criteria, procedures, posted notifications, and escape route exit signs.

9.3.7 Compensation Access Control

A compensation control is an alternative either to use instead of when primary control or to increase the effectiveness of primary control. For example, the security policy of a company may make it mandatory for the use of smartcards by all employees, but sometimes the arrangement of a smartcard for new employees gets delayed. In that case, as a compensating control, the organization could issue hardware tokens to employees. These tokens are stronger authentication than a simple username and password.

According to the implementation of access controls, it can be categorized into administrative, logical/technical, or physical control. These types can have other examples of access control types mentioned previously.

9.3.8 Administrative Access Controls

Administrative or management access controls include the procedures and policies instructed by an organization's security policy and other requirements or regulations. Mostly, these controls cover business and personnel practices.

Examples of administrative access controls involve policies, hiring practices, procedures, background checks, testing, security awareness and training efforts, classifying and labeling data, reports and reviews, and personnel controls.

9.3.9 Logical/Technical Controls

Logical/Technical access controls involve the software or hardware mechanisms used to manage access and provide protection for systems and resources. It is all about technology, as it implies.

Examples of technical or logical access controls involve authentication methods like usernames, biometrics, passwords, scanners, smartcards, encryption, routers, firewalls, protocols, constrained interfaces, access control lists, clipping levels, and intrusion detection systems.

9.3.10 Physical Access Controls

Physical access controls enlist those items that can be physically touched. These cover physical mechanisms worked to monitor, prevent, or detect direct contact with areas or systems within a facility.

Examples of physical controls cover lights, fences, guards, video cameras, alarms, motion detectors, locked doors, sealed windows, cable protection, swipe cards, badges, guard dogs, mantraps, laptop locks, and many more (Fig. 9.1).

Assignment Question

1.   What are all the factors which are used for managing access to assess?
2.   What is the purpose of using access control strategies?
3.   What do you mean by access control models?
4.   Detail about preventive access control
5.   Brief about the importance of compensation access control and its application.

Fig. 9.1 Defense-in-depth
implementation in security
controls

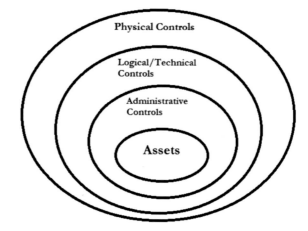

Multiple Choices Question

1. Which management provides a single point of access for end-users to manage
    their identity information and a management console for administrators?

    (A) Identity (B) User (C) A and B

2. The effectiveness of access controls is assessed by access

    (A) Review audits (B) Security audits (C) None of the above

3. Who gains illegal access to a computer system is known as……..

    (A) Hacker (B) Worm (C) Pirate (D) Thief

4. ……..Two main types of access control lists.

    (A) IEEE (B) Extended (C) Standard (D) Specialized

5. Which of the following refers to the violation of the principle if a computer is
    no more accessible?

    (A) Access control (B) Confidentiality (C) Availability (D) All of the above

Answers

1. Identify
2. An exit conference
3. Hacker
4. Extended
5. **Availability**.

Summary Question

1. What do you mean by physical access control and its strategies?
2. Detail explanation of physical access control and its limitations
3. What do you mean by technical and logical controls?
4. Why is access control needed?
5. List the application of access control strategies in assets.

Chapter 10
Manage the Identification and Authentication of People, Devices, and Services

The process of a subject claiming an identity is known as identification. To start the authentication, authorization, and accountability processes, the subject needs to provide an identity to a system. Typing a username, waving a token device, speaking a phrase, or positioning your hand, finger, or face in front of a camera or the range of a scanning device can be the process of providing identity.

For authentication, all subjects must have unique identities. Authentication compares one or more factors against a database of valid identities, such as user accounts, to verify the identity of the subject. The information the subjects use to verify themselves in the authentication process is private and needs to be protected. As an example, passwords are rarely stored in clear text within a database. Instead, authentication systems store hashes of passwords within the authentication database. The level of security of the authentication system is reflected by its and the subjects' ability to maintain the secrecy of the authentication information.

Identification and authentication always occur together as a single two-step process. In this two-step process, identity is the first step, and providing authentication information is the second step. A subject can gain access to a system only by completing both of these steps. Each authentication technique or factor has unique benefits and drawbacks. That's why it is important to evaluate each mechanism in the context of the environment where it will be deployed. For example, a facility that processes Top secret materials require very strong authentication mechanisms. In contrast, authentication requirements within a classroom environment are significantly less.

The combination of a username and a password is a simple example of the identification and authentication process. To identify themselves, users use their usernames, and for the authentication process, they provide their passwords. Of course, there are many more identification and authentication methods, but this simplification helps keep the terms clear.

B. S. Rawal et al., *Cybersecurity and Identity Access Management*,
https://doi.org/10.1007/978-981-19-2658-7_10

10.1   Registration and Identity Proofing

A user needs to go through the registration process when he or she is given an identity for the first time. Within an organization, new employees provide appropriate documents to prove their identity during the hiring process. Then the employees of the Human Resource (HR) department begin the process of creating their user ID.

The more secure the authentication process is, the more complex the registration process gets. For example, if the organization uses fingerprinting as a biometric method for authentication, registration includes capturing user fingerprints.

Identity proofing has different functionality. It is used when users interact with online sites, for example, an online banking site. While creating an account, the bank will take some steps to validate the user's identity. Usually, they will ask for information that is only known to the user and the bank, such as account numbers and personal information about the user, such as a national identification number or Social Security Number. Also, during the initial registration process, the bank will ask the user to provide some additional information such as the middle name of his/her mother, the user's favorite color, or the model of their first car. Later, if the user wants to change their password or transfer money, the bank may use these questions for identity proofing.

10.2   Authorization and Accountability

Authorization and accountability are the two additional security elements in an access control system.

Authorization

Authorization is the process of granting access to subjects based on proven identities. It indicates who is allowed to perform certain operations. If the action is allowed, the subject is authorized; if disallowed, the subject is not authorized. For example, if an employee of an organization attempts to open a file, the authorization system will check if that employee at least has permission to read it. It is important to understand that just because the user can authenticate to a system, that doesn't mean he can open and overwrite any file on the system. Instead, users are authorized to access specific objects based on their proven identity. The process of authorization ensures that the requested activity or object access is possible based on the privileges assigned to the subject.

Identification and authentication are "all-or-nothing" aspects of access control. Either a user's credentials prove a professed identity, or they don't. In contrast, authorization can have a wide range of variations. For example, a user may be able to view a file but not edit or delete it.

Accountability

To make users and other subjects accountable for their actions, auditing needs to be implemented. Auditing is the process of tracking and recording subject activities within logs. Logs typically record who took a certain action, what the action was, and when and where the action was taken. One or more logs create an audit trail that researchers can use to reconstruct events and identify security incidents. When investigators review the contents of audit trails, they can provide evidence to hold people accountable for their actions.

There is an important point to mention about accountability. Effective identification and authentication are the only two factors accountability rely on. It doesn't require effective authorization. In other words, after identifying and authenticating users, accountability mechanisms such as audit logs can track their activity, even when they try to access resources that they aren't authorized to access.

Authentication Factors

Authentication factors are security credentials that are used to verify the identity and authorization of a user attempting to gain access, send communications, or request data from a secured network, system, or application.

Organizations used to rely on unique usernames and user-selected passwords as the primary method of authenticating user identity and providing access to systems. But now, as they are placing more and more emphasis on security due to regulatory and compliance concerns, many organizations use multiple authentication factors to control access to secure systems and applications.

There are three basic methods of authentication, which are also known as factors. They are as follows:

1. Something you know: Examples include a password, personal identification number (PIN), or passphrase.
2. Something you have: Physical devices that a user possesses can help them provide authentication. Examples include a smartcard, hardware token, smartcard, memory card, or USB drive.
3. Something you are: It is a physical characteristic of a person identified with different types of biometrics. Examples in the something-you-are category include fingerprints, voiceprints, retina patterns, iris patterns, face shapes, palm topology, and hand geometry. Examples in the something-you-do category include signature and keystroke dynamics, also known as behavioral biometrics.

When implemented correctly, the something-you-are authentication factor is the strongest while the something-you-know is the weakest one. In other words, passwords are the weakest, and a fingerprint is stronger than a password. However, there some ways that attackers can use to bypass even the strongest authentication factor. For example, an attacker may be able to create a duplicate fingerprint on a gummi bear candy and fool a fingerprint reader.

Apart from these three basic factors, there is another factor known as somewhere-you-are, which is used sometimes. Based on a specific computer, a phone number identified by caller ID, or a geographic location identified by an IP address, it can identify a subject's location. To control access by using physical location requires the subject to be present in a specific location. For example, consider remote access users who dial in from their homes. Caller ID and callback techniques can verify that the user is calling from home.

As a dedicated attacker can spoof any type of address information, this factor to use on its own. However, using it in combination with other factors can be very effective.

Passwords

The use of a password is the most common authentication technique. It's a string of characters entered by a user. Passwords are typically static, which means they stay the same for a certain period. The problem with using a static password as an authentication technique is that it is the weakest form of authentication. Here are some of the reasons for passwords being a weak security mechanism:

- People often set a password that they can easily remember. This makes the password easy to guess or crack.
- Many users write down their passwords if the passwords are randomly generated and hard to remember.
- Users often share their passwords or forget them.
- Attackers detect passwords through many means, including observation, sniffing networks, and
- stealing security databases.
- Sometimes user passwords are transmitted without or easy to break encryption protocols.
- On many occasions, publicly accessible online locations are used to store the password database.
- Weak passwords are greatly vulnerable to brute-force attacks.

10.3 Effective Password Mechanisms

Setting Strong Passwords

Passwords can be an effective authentication technique if users use strong passwords. A strong password is sufficiently long and has a mixture of upper case, lower case, numbers, and special characters. To ensure users use strong passwords, organizations often include a written password policy in the overall security policy. Then the policy is enforced with technical controls by the IT security professionals of the organization. Here are some of the common password policy settings:

- Forcing users to change their passwords periodically, such as every 60 days.
- Allowing only complex passwords. Because an eight-character password containing uppercase characters, lowercase characters, symbols, and numbers is much stronger than an eight-character password using only numbers.

- Ensuring that user passwords are of a certain length. Length The length is the number of characters in the password. Shorter passwords are easier to crack. For example, a complex five-character password can be cracked using a password cracker within a second. In contrast, it takes thousands of years to crack a complex 12-character password. Many organizations require privileged account passwords to be at least 15 characters long.
- Not allowing users to reuse one of their previously used passwords. A password history remembers a certain number of previous passwords and prevents users from reusing a password in history. A minimum password age setting is often combined with it to prevent users from changing a password repeatedly until they can set the password back to the original one.

The main reason users often don't set a strong password is that they don't understand its importance. Even if they do understand its importance, in many cases, they don't know how to create a strong password that they can easily remember. These suggestions listed below will help them with that:

- Avoid using any part of your email address, name, employee number, national identification number or Social Security Number, phone number, extension, or other identifying name or code.
- Do not use the information available from social network profiles such as a family member's name, a pet's name, or your birth date.
- Use nonstandard capitalization and spelling.
- Use special characters and numbers to replace letters.

In many cases, systems automatically create the initial passwords for a user. Usually, the generated password contains two or more unrelated words joined together with a number or symbol in between. Systems can easily generate these composition passwords. However, users should not use them for an extended period as they are vulnerable to password-guessing attacks.

Password Phrases

A passphrase is more secured than a basic password. A passphrase is a string of characters similar to a password, but that has a unique meaning to the user. Users usually use basic sentences that they can remember as passphrase after slightly modifying them with special characters and numbers. For example, "I am CISSP certified" can be converted to "I@mCI$$Pcertified." Using a passphrase has several benefits. It is difficult to crack a passphrase using a brute-force tool, and it encourages the use of a lengthy string with numerous characters, but it is still easy to remember.

Cognitive Passwords

Cognitive passwords is another password mechanism. It is a series of questions about facts or predefined responses that only the subject should know. During the initial registration of the account, authentication systems often collect the answers to these questions, but they can be collected or modified later. For example, the user might be asked a few questions, such as:

- What is the name of your best friend?
- What is the name of your first pet?
- What is the color of your first car?
- What is the middle name of your mother?

Later, the system uses these questions for authentication. If the user answers all the questions correctly, the system authenticates the user. The most effective cognitive password systems collect answers for several questions and ask a different set of questions each time they are used. Cognitive passwords are often used for password management like self-service password reset systems or assisted password reset systems. For example, if a user forgets his password and asks for help, the system challenges him with the cognitive password questions. If the user can give correct answers, the system allows him to set a new password.

Smartcards and Tokens

Smartcards and hardware token are both examples of a something-you-have authentication factor. They are usually used in combination with other factors of authentication, providing multifactor authentication.

Smartcards

A smartcard is an ID or a badge that looks quite similar to a credit card. It has an integrated circuit chip embedded in it that contains information about the authorized user. This information is used for identification and/or authentication purposes. Nowadays, many smartcards have a microprocessor built-in and include one or more certificates. The certificates are used for asymmetric cryptography, such as encrypting data or digitally signing an email. So, smartcards provide an easy way to carry and use complex encryption keys while being tamper resistant.

However, as the smartcard of one user can be used by anyone else, it isn't an effective identification method on its own. That's why in most cases, users are required to use one or more other authentication factors such as a PIN or a combination of a username and a password with it.

Tokens

A token is a device that generates passwords, and the user can carry them with him. A common token used today includes a display that shows a six- to eight-digit number. An authentication server stores the details of the token, so at any moment, the server knows what number is displayed on the user's token. Tokens are usually used together with another authentication mechanism. For example, the user may enter a pin and then enter the number displayed on the token. Tokens generate a dynamic one-time password. Therefore, they are more effective than a static password.

Hardware tokens can be of two types:

- Synchronous dynamic password tokens
- Asynchronous dynamic password tokens.

Biometrics

Biometrics is another widely used identification and authentication technique. It falls into the something-you-are authentication category. Instead of a username or account ID, using a biometric factor for the identification process requires a one-to-many search of the offered biometric pattern against a stored database of enrolled and authorized patterns. Capturing a single image of a person and searching a database of many people looking for a match is an example of a one-to-many search. Whereas using a biometric factor as an authentication technique requires a one-to-one match of the offered biometric pattern against a stored pattern for the offered subject identity. In other words, the users want to access a system by claiming an identity and the system then checks if the person matches the claimed identity. Physical access controls usually have biometric factors installed as an identification technique.

Biometrics characteristics can be of two types: physiological or behavioral. Physiological biometric methods include face scans, fingerprints, retina scans, iris scans, palm scans, hand geometry, and voice patterns. Behavioral biometric methods include signature dynamics and keystroke patterns. Behavioral biometrics are sometimes referred to as something-you-do authentication.

Multifactor Authentication

Multifactor authentication is an authentication process where two or more factors are used. Two-factor authentication requires two different factors to provide authentication. For example, if you want to withdraw money from an ATM, you need to insert your ATM card and then enter a pin. Similarly, smartcards typically require users to insert their card into a reader and also enter a PIN. As a general rule, using more types or factors results in more secure authentication. Multifactor authentication must use multiple types of factors, such as the something-you-have and something-you-know factor. However, an authentication process requiring users to enter a password and a PIN is not a multifactor authentication as both methods are from something-you-know factors. Using two authentication methods of the same factor isn't more secured than it would be if just one method were used because the same attack that could steal or obtain one could also obtain the other.

Device Authentication

In the past, they are logging into a network that required users to use a company-owned system such as a desktop PC. For example, in a Windows domain, the user computer joins the domain and has computer accounts and passwords similar to user accounts and passwords. If the computer hasn't joined the domain, or its credentials are out of sync with a domain controller, users cannot log on from this computer. But nowadays, almost all employees bring their own devices to work and hook them up to the network. Although some organizations allow these, they implement BYOD security policies as a measure of control. These devices aren't necessarily able to join a domain, but it is possible to implement device identification and authentication methods for these devices. One method is device fingerprinting. Users can register their devices with the organization and associate them with their user accounts. A device authentication system captures the device characteristics during registration. For this process, a user needs to access a web page with the device. The registration system then uses characteristics such as the web browser, browser fonts, browser plug-ins, operating system and version, time zone, data storage, screen resolution, cookie settings, and HTTP headers to identify the device. If the user tries to log in from the device, the authentication system checks the user account for a registered device. It then verifies the characteristics of the user's device with the registered device. Although some of these characteristics may change over time, this process has proven to be a successful device authentication method.

Assignment Questions

1. What do you mean by registration and identifying proofing?
2. What is the purpose of using Authorization and accountability?
3. List the security benefits of firewall
4. List the authorization methods used in proofing
5. What is the significance of registration in identifying proofing?

Multiple Choices Question

1. In cryptography, the shuffling of letters is named as………..

   (A) trans-positional approach (B) substitution approach (C) both A and B (D) Quadratic ciphers
2. A computer……… is a malicious code which self-replicates by copying itself to other programs.

   (A) **Virus** (B) Application (C) Worm (D) Program
3. ………………… is the developer of this ticket-based authentication system.

   (A) IBM (B) TCS (C) SPLM
4. ……………….. is a system that allows the creator, owner, or data custodian of an object to define and control access to that object.

   (A) DACs, (B) TBAC, (C) RBAC
5. Who should have a single resource for common operations?

   (A) Users (B) Admin (C) Enrollment

Answers

1. Trans-positional approach
2. **Virus**
3. IBM
4. DACs
5. Users.

Summary Questions

1. What are all the methods which are used in firewall protection and security
2. What are all models used in security protocols for data processing?
3. List the limitation of firewalls in internet security architecture.
4. What do you mean by symmetric cryptography technique?
5. What are all the limitations used in the security architecture of the firewall process?

Chapter 11
Integrate Identity as a Third-Party Service

11.1 Identity Management Techniques

Identity management techniques can be divided into two categories:

1. Centralized
2. Decentralized.

In a centralized identity management technique, the authorization process is performed by a single entity, whereas various entities located throughout a system perform the authorization process of a decentralized identity management technique. No matter where the centralized or decentralized access control techniques are applied, their benefits and drawbacks stay the same.

An individual or a small group of people can manage a centralized access control system. Having a small number of people working on this reduces the administrative expense. However, the working team needs to carefully implement changes in a centralized access control system as a single change can affect the entire system.

In contrast, in a decentralized access control system, any change made to any individual access control point needs to be repeated at every other point. That's because the system is divided into multiple access control points, and several teams or multiples individuals are in charge of administrating them. Having multiple teams or individuals working on different access control points of a system makes maintaining consistency across a system difficult as well as increases administrative expenses.

Single Sign-on

Single sign-on (SSO) is a centralized access control technique that allows a subject to access multiple resources on a system by only getting authenticated once. For example, if a user wants to connect to a network and the system gives him access after authenticating, then he will be able to access resources throughout the network without being prompted to authenticate again.

SSO is not only user-friendly but also increases security. When users need to use multiple usernames and passwords to get access to authorized resources, they are

© The Author(s), under exclusive license to Springer Nature Singapore Pte Ltd. 2023
B. S. Rawal et al., *Cybersecurity and Identity Access Management*,
https://doi.org/10.1007/978-981-19-2658-7_11

most likely to write them down. This ultimately weakens security. But if they have to remember just one password, they are unlikely to write it down. SSO also eases administration by reducing the number of accounts required for a subject.

The main disadvantage of using SSO is that once an account is compromised, an attacker gains unrestricted access to all of the authorized resources. However, most SSO systems include methods to protect user credentials.

The following sections discuss several common SSO mechanisms.

LDAP and Centralized Access Control

Usually, a centralized access control system is used within a single organization. For example, a directory service that includes information about subjects and objects is a centralized database. Directory services are often based on the Lightweight Directory Access Protocol (LDAP). For example, the Microsoft Active Directory Domain Services is LDAP based.

To understand the LDAP directory, you can think of it as a telephone directory for network services and assets. Users, clients, and processes can search the directory service to find where a desired system or resource resides. Before subjects can perform queries and lookup activities, they must authenticate to the delivery service. Once the authentication is completed, the directory service will allow the subject to see certain information based on that subject's assigned privileges. Multiple domains and trusts are commonly used in access control systems. A security domain is a collection of subjects and objects that share a common security policy, and individual domains can operate separately from other domains. The function of trust is to create a security bridge and allow users to access resources from one domain to another. Trusts can be one way only, or they can be two ways.

LDAP and PKIs

Public key infrastructure (PKI) is a group of technologies used to manage digital certificates during the certificate life cycle. It uses LDAP when integrating digital certificates into transmissions. Clients often need to query a certificate authority (CA) for information on a certificate, and LDAP is one of the protocols used.

To support single sign-on capabilities, centralized access control and LDAP can be used.

Kerberos

Ticket authentication is a mechanism that employs a third-party entity to prove identification and provide authentication. Kerberos is the most common and well-known ticket system. The current version of Kerberos is Kerberos V5, which is based on the Kerberos authentication system developed at MIT. Under Kerberos, a client (generally either a user or a service) sends a request for a ticket to the key distribution center (KDC). The KDC creates a ticket-granting ticket (TGT) for the client, encrypts it using the client's password as the key, and sends the encrypted TGT back to the client. The client then attempts to decrypt the TGT using its password. If the client successfully decrypts the TGT (i.e., if the client gave the correct password), it keeps the decrypted TGT, which indicates proof of the client's identity. The TGT, which expires at a specified time, permits the client to obtain additional tickets, which permit specific services. The requesting and granting of these additional tickets are user transparent.

Kerberos V5 relies on symmetric key cryptography using the Advanced Encryption Standard (AES) symmetric encryption protocol. There are several different elements used by Kerberos V5 that are important to understand:

Key Distribution Center

The key distribution center (KDC) is the authentication service providing a third party. To authenticate clients to servers, Kerberos uses symmetric key cryptography. KDC has all clients and serves registered to it and maintains the secret keys for all network members.

Kerberos Authentication Server

The functions of KDC that are a ticket-granting service (TGS) and an authentication service (AS) are hosted by the authentication server. However, a different server can be used to host the ticket-granting service. The authentication service verifies or rejects the authenticity and timeliness of tickets.

Ticket-granting Ticket

A ticket-granting ticket (TGT) provides proof that a subject has authenticated through a KDC and is authorized to request tickets to access other objects. A TGT is encrypted, and an asymmetric key, an expiration time, and the user's IP address are included in it. When subjects request tickets to access objects, they present the TGT.

Ticket

A ticket, which is also called a service ticket (ST), is an encrypted message that provides proof that a subject is authorized to access an object. Subjects request tickets to access objects. Kerberos then checks whether they have the authentication and authorization to access the object. If they have authenticated and are authorized, Kerberos issues them a ticket. Kerberos tickets have specific lifetimes and usage parameters. Once a ticket expires, a client must request a renewal or a new ticket to continue communications with any server.

To operate, Kerberos requires a database account that is usually contained in a directory service. It uses an exchange of tickets between clients, network servers, and the KDC to prove identity and provide authentication. This way, both the client and server get the assurances of the identity of the other when a client request resources from the server. These encrypted tickets also ensure that login credentials, session keys, and authentication messages are never transmitted in cleartext.

Kerberos Logon Process

The Kerberos login process works as follows:

- At first, the user types a username and password into the client
- The client uses AES to encrypt the username for transmission to the KDC

- The KDC uses a database of known credentials to verify the username
- An asymmetric key encrypted with a hash of the user's password is generated by the KDC. Both the client and the Kerberos servers use this symmetric key. The KDC also generates an encrypted time-stamped TGT
- The KDC then transmits the encrypted symmetric key and the encrypted time-stamped TGT to the client
- The client installs the TGT for use until it expires. The client also decrypts the symmetric key using a hash of the user's password.

Accessing Objects Through Kerberos Server

When a client wants to access an object, such as a resource hosted on the network, it must request a ticket through the Kerberos server. The following steps are involved in this process:

- The client sends its TGT back to the KDC with a request for access to the resource
- The KDC checks the validity of the request and its access control matrix to verify whether the user has sufficient privileges to access the requested resource
- The KDC generates a service ticket and sends it to the client
- The client sends the ticket to the server or service hosting the resource
- The server or service hosting the resource verifies the validity of the ticket with the KDC
- Kerberos activity is completed upon the verification of identity and authorization. A session is then opened by the server or service host with the client to begin communications or data transmission.

Kerberos is a versatile authentication mechanism that works over local LANs, remote access, and client-server resource requests. However, the main drawback of using Kerberos is that its KDC can work as a single point of failure. If the KDC is compromised, the secret key for every system on the network is also compromised. Also, subject authentication can't occur if a KDC goes offline. Because of having strict time requirements and the default configuration requires that all systems be time-synchronized within five minutes of each other. A previously issued TGT will no longer be valid, and the system will not be able to receive any new tickets if a system is not synchronized or the time is changed. As a result, the client won't be able to access any protected network resources.

Federated Identity Management and SSO

SSO is being used on internal networks for quite a while, but not on the internet. However, the demand for an SSO solution for users accessing resources over the internet has increased with the drastic increase of cloud-based applications. Federated identity management is a form of SSO that meets this need.

The management of user identities and their credentials is known as identity management. Federated identity management has taken this one step further by enabling multiple organizations to join a federation. When multiple organizations join a federation, they agree on a method to share identities between them. Users in each organization can log on once in their organization, and their credentials are matched with federated identity. By using this federated identity, they can then access resources in any other organization within the group.

A federation can be composed of multiple unrelated networks within a single university campus, multiple colleges and university campuses, multiple organizations sharing resources, or any other group that can agree on a common federated identity management system. Members of the federation match user identities within an organization to federated identities. For example, federated SSO systems are used by many corporate online training websites. When the organization coordinates with the online training company for employee access, they also coordinate the details needed for federated access. A common method is to match the user's internal login ID with federated identity. Users log on within the organization using their normal login ID. When the users access the training website with a web browser, the federated identity management system uses their login ID to retrieve the matching federated identity. The user is authorized to access the web pages granted to the federated identity if it finds a match. These details are managed by administrators behind the scenes. However, the process is usually transparent to users.

One of the challenges that come up when multiple companies communicate in a federation is finding a common language. They often have different operating systems, but they still need to share a common language. Federated identity systems often use the Security Assertion Markup Language (SAML) and/or the Service Provisioning Markup Language (SPML) to meet this need.

More Examples of Single Sign-on

Kerberos is the most recognized and deployed form of single sign-on within an organization. However, there are some other SSO mechanisms that you may encounter. We will be discussing briefly some of those in this section.

Scripted Access or Logon Scripts

It uses an automated process to transmit login credentials at the start of a login session to establish communication links. Even if an environment requires a unique authentication process to connect to each server or resource, scripted access can often simulate SSO. It is great to implement SSO in environments where true SSO technologies are not available. As the scripts and batch files usually contain access credentials in clear text, they should be stored in a secured area.

The Secure European System for Applications in a Multivendor Environment (SESAME).

It was initially developed to solve the weaknesses in Kerberos. However, this ticket-based authentication system could not solve all the problems with Kerberos. Eventually, newer Kerberos versions and various vendor implementations resolved the initial problems with Kerberos, bypassing SESAME. That is the reason why SESAME is no longer considered a viable product in today's professional security world.

KryptoKnight

IBM is the developer of this ticket-based authentication system. There are some similarities between KryptoKnight and Kerberos. But, KryptoKnight uses peer-to-peer authentication instead of a third party. It was incorporated into the NetSP product. Like SESAME, KryptoKnight and NetSP never took off and are no longer widely used.

OAuth and OpenID

OAuth and OpenID are the two newer examples of SSO used on the internet. OAuth is an open standard designed to work with HTTP, and it allows users to log on with one account. For example, users can log onto their Google account and use the same account to access Facebook and Twitter pages. Google supports OAuth 2.0, which is not backward compatible with OAuth 1.0. OpenID is also an open standard. But rather than as an IETF RFC standard, OpenID Foundation maintains it. OpenID can be used in conjunction with OAuth or on its own.

11.2   Credential Management Systems

A credential management system is a way of storing user credentials when SSO isn't available. Users can store credentials for websites and network resources that require a different set of credentials. The credentials are secured with encryption by the management system to prevent unauthorized access. The credential manager tool of the windows systems is an example of it. Users enter their credentials into the Credential Manager, and when necessary, the operating system retrieves the user's credentials and automatically submits them. If users use this to visit a website and enter the username and password, later, when they revisit the website, the Credential Manager automatically recognizes the URL and provides the credentials. Also, there are some third-party credential management systems like KeePass available for users to use.

Identity as a Service

Identity as a Service or IDaaS is cloud-based authentication built and operated by a third-party provider. Enterprises that subscribe to IDaaS companies are provided with cloud-based authentication or identity management.

The X-as-a-service model in information technology is easy to understand. It means a third-party provider serves a company some features through a remote connection, as opposed to the features being managed by the company's employees. We can use local emails such as Microsoft Outlook or Thunderbird as examples. They operate primarily on one's computer versus cloud email, such as Gmail, being provided to users as a service through web connections. Identity, security, and other features can similarly be provided as a service.

IDaaS effectively provides SSO for the cloud and is especially useful when internal clients access cloud-based Software as a Service (SaaS) applications. The goal of an Identity Service is to ensure users are who they claim to be and to give them the right kinds of access to software applications, files, or other resources at the right times. If a company decides to build and operate the infrastructure to make this happen by itself, then the company has to figure out what to do every time a problem comes up. If Bring Your Device (BYOD) employees are changing to different types of phones, for example, the local identity provisioning has to adapt immediately. It is much simpler to implement a centralized cloud-based system created by identity experts who have already solved such problems for hundreds of organizations.

Functionality of IDaaS

Basic functionality of IDaaS includes:

Access: It includes single sign-on (SSO), authorization enforcement, and user authentication. Employees, customers, and partners can get fast, easy, and secure access to all Software as a Service, mobile, and enterprise apps with single authentication from corporate credentials by using single sign-on (SSO). For authentication, different adaptive methods can be used based upon the level of risk, changes in the situation, or sensitivity of the application.

Identity Governance and Administration (IGA): It provides target applications with individual service identities by managing identity ad access life cycles across several systems. Core functions of IGA include Identity Lifecycle Management, Access Requests, Role and Policy Management, Workflow Orchestration, Auditing, Password Management, Reporting, and Analytics, Access Certification.

Intelligence: It provides reports related to logs and answers log related questions. User data can be synced with enterprise and web applications by using on-premises provisioning.

Assignment Questions

1. What is a centralized identity management technique?
2. Describe single sign-on
3. Give the definition of Lightweight Directory Access Protocol
4. How the centralized access control and LDAP can be used to support single sign-on capabilities?
5. What is the use of ticket-granting services?

Multiple Choice Questions

1. .... is the authentication service providing a third party
   (A) Key Distribution Center (B) Kerberos Authentication Server (C) **A and B**
2. Who checks the validity of the request and its access control matrix to verify whether the user has sufficient privileges to access the requested resource?
   (A) TGT (B) KDC (C) SAML
3. Federated identity systems often use.............................
   (A) Security Assertion Markup Language (SAML)

(B) Service Provisioning Markup Language (SPML)

(C) A and B

4. ………………….. is the developer of this ticket-based authentication system

(A) IBM (B) TCS (C) SPLM

5. Which management system is a way of storing user credentials when SSO isn't available?

(A) Identity (B) Credential (C) Kerberos

Answers

1. Key Distribution Center
2. KDC
3. A and B
4. IBM
5. Credential

Summary Questions

1. What is the difference between KDC and TGT?
2. What is symmetric key cryptography?
3. Mention the use of federated identity systems
4. What is single sign-on?
5. What is KryptoKnight?

Chapter 12
Implement and Manage Authorization Mechanisms

The access control method used by the IT system determines the method of authorizing subjects to access objects. There are varieties of categories for access control mechanisms, and the CISSP CIB specifically mentions four among two discreet groups:

1. Discretionary access control (DAC)
2. Non-discretionary access control (DAC): Mandatory Access Control (MAC), role-based access control (role-BAC), and role-based access control (rule-BAC).

Discretionary Access Controls

Discretionary access controls (DACs) are a system that allows the creator, owner, or data custodian of an object to define and control access to that object. All objects have specific owners, and the decision or discretion of the owner determines the variety of access control. For example, a user owns a spreadsheet if he is the creator of that new one. As an owner, the user has the full right to modify the permissions of the file to deny or grant access to other users. DAC has an access control subset based on identity. Users are identified in the system based on their own identity, and resource ownership is assigned to identities.

Access control lists (ACLs) implement a DAC model on objects. Each ACL is responsible for the denial and grant of access to the subject. A centrally controlled management system is not offered by ACLs because the ACLs can be altered on owners' objects at will. Access to objects can easily be changed, especially in comparison with the static nature of Mandatory Access Controls (MACs). Administrators can easily suspend user privileges within a DAC environment while they are away, like on vacation. Also, when a user leaves an organization, it's easy to disable his accounts.

Non-discretionary Access Controls

Discretionary and non-discretionary access controls differ in how they are controlled and managed. Non-discretionary access controls are centrally administered by

© The Author(s), under exclusive license to Springer Nature Singapore Pte Ltd. 2023 167
B. S. Rawal et al., *Cybersecurity and Identity Access Management*,
https://doi.org/10.1007/978-981-19-2658-7_12

administers, and slight changes in it affect the entire environment. On the contrary, owners are responsible for the change in their objects in discretionary access control models, and the other parts of the environment aren't affected by their changes. In a non-DAC model, user identity isn't focused on access. Rather, access is managed by a static set of rules governing the whole environment. Though non-DAC systems are less flexible, they are easier to manage as they are centrally controlled. In general, all the models other than a discretionary model are non-discretionary models, including role-based, rule-based, and lattice-based access controls.

Role-Based Access Control

Systems that employ task-based or role-based access controls define a subject's ability to access an object based on the subject's assigned tasks or role. Most often, subjects implement role-based access control (role-BAC) using groups. For example, a bank usually consists of loan officers, tellers, and managers where administrators can create a group named Loan Officers and place each loan officers' user accounts into that group and then assign appropriate facilities to the group, as shown in Fig. 12.1. Administrators simply add the new loan officer's account into the Loan Officers group if the organization hires one. The new employee will have all the same privileges as other loan officers in this group automatically. Administrators can follow similar steps for tellers and managers.

This helps the prevention of privilege creep by enforcing the principle of least privilege. The tendency for privileges to accrue to users over time when roles and access needs change is called privilege creep. Ideally, when users change jobs within an organization, administrators revoke user privileges. However, when privileges are

Fig. 12.1 Role-based access controls

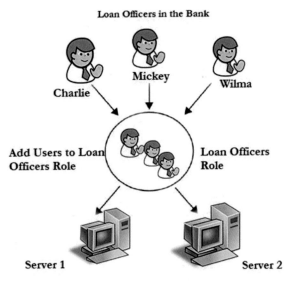

directly assigned to users, it is hard to identify and revoke all of a user's unnecessary privileges.

Administrators can simply remove the user's account from a group to easily revoke unnecessary privileges. A user no longer has the privileges assigned to a group as soon as an administrator removes the user from the group. For example, if a loan officer is assigned to another department, the loan officer's account can simply be removed from the Loan Officers group and added to the new departments' group by administrators. This immediately revokes all the Loan Officers group privileges from the user's account and assigns them to the privileges of another group.

Job descriptions or work functions identify roles (and groups) to administrators. In many cases, this follows the organizational chart containing the organization's hierarchy documented in it. As per hierarchy, users in a temporary job have lesser access to resources than users who occupy management positions. Administrators can easily grant multiple permissions simply by adding a new user into the appropriate role in role-based access controls, which are very useful in dynamic environments with frequent personnel changes. In this process, multiple roles or groups perform by the same user worth noting. For example, using the same bank scenario, managers might perform at the same time the Managers role, the Loan Officers role, and the Tellers role. In this way, managers can access all of the same privileges that their employees can access.

DAC and role-BAC both use groups to organize users into manageable units, which is a bit confusing, but there is a difference in deployment and use. In the DAC model, a specific owner determines who has access to his objects. On the other hand, in a role-BAC model, appropriate subject privileges are assigned to roles or groups determined by administrators. But in a strict role-BAC model, users are not assigned privileges directly; rather, administrators only grant privileges by adding user accounts to roles or groups.

The task-based access control (TBAC) is quite similar to role-BAC. But there is a slight difference. Here each user, instead of being assigned to one or more roles, is assigned to an array of tasks. A person associated with a user account is assigned for all these items related to work tasks. Under TBAC, rather than user identity, the focus is on controlling access by assigned tasks.

Rule-Based Access Controls

A rule-based access control (rule-BAC) uses a set of restrictions, filters, or rules to determine what can occur on a system and cannot. Like it includes granting a subject whether it can access an object and able to act or not. A specific characteristic of rule-BAC models is that their global rules apply to all subjects.

For example, a firewall is a common model for a rule-BAC model. Firewalls form a set of filters or rules within an ACL, defined by an administrator. The firewall only allows traffic to meet one of the rules by examining all the traffic going through it. Firewalls include a final rule denying all other traffic, often referred to as the implicit deny rule. For example, the last rule might be "deny all," which orders the firewall to block all other traffic in or out of the network. On the other hand, it occurs when the traffic didn't meet the condition of any previously explicitly defined rule, then the traffic is blocked by the final rule. Sometimes you can see this final rule in the ACL. Other times, the deny rule is implied implicitly as the final rule, but the ACL does not explicitly announce this.

Attribute-Based Access Controls

Traditional rule-BAC models are applied to all users as it contains global rules. But an advanced implementation of a rule-BAC is known as the attribute-based access control (ABAC) model. Multiple attributes for rules are included in the policies of ABAC models. Applying ABAC models, the networking of any software can be defined.

As an example, a software-defined wide area network (SD-WAN) solution has been created by CloudGenix that implements policies to allow or block traffic. Plain language statements such as "Allow Managers to access the WAN using tablets or smartphones" are created by administers using ABAC policy, allowing Managers to access the WAN using smartphones or tablet devices. The rule-BAC is universal, whereas ABAC is much more specific, improving the rule-BAC model.

Mandatory Access Controls

Mandatory Access Control (MAC) is the strictest of all levels of control. The government defined the design of MAC and primarily used it.

To control access to resources, MAC takes a hierarchical approach. Under a MAC enforced environment, access to all resource objects such as data files is controlled by settings defined by the system administrator. The operating system, based on system administrator configured settings, strictly controls all access to resource objects. The users can't change the access control of a resource under MAC enforcement. Mandatory Access Control begins with security labels assigned to all resource objects on the system. These security labels contain two pieces of information—a classification (top secret, confidential, etc.) and a category (which is essentially an indication of the management level, department, or project to which the object is available).

Similarly, each user account on the system also has classification and category properties from the same set of properties applied to the resource objects. When a user attempts to access a resource under Mandatory Access Control, the operating system checks the user's classification and categories and compares them to the properties of the object's security label. If the user's credentials match the MAC security label properties of the object, access is allowed. It is important to note that both the classification and categories must match. A user with top secret classification, for example, cannot access a resource if they are not also a member of one of the required categories for that object.

Mandatory Access Control is by far the most secure access control environment but does not come without a price. Firstly, a considerable amount of planning is required before implementing MAC. Once implemented, it needs to constantly update object and account labels to accommodate new data, new users, and changes in the categorization and classification of existing users that result in a high system management overhead.

A hierarchy of sensitivity is indicated by security classifications. For example, in military security labels of Top Secret, Secret, Confidential, and Unclassified, the Top Secret label contains the most sensitive data, whereas unclassified contains the least sensitive one. This hierarchy cleared someone for Top Secret data as well as for Secret and less sensitive data. However, classifications don't need lower levels. It is better to use MAC labels if someone with higher-level label access does not need clearance for a lower-level label. A distinguishing point about the MAC model is that here every object and every subject is labeled. Here the system makes an authorization of access based on predefined assigned labels.

The following three types of environments are used as classifications within a MAC model:

Hierarchical Environment

A hierarchical environment in an ordered structure contains various classification labels from low security to medium security to high security, such as Confidential, Secret, and Top Secret, respectively. There is a relationship between each classification or level label in the structure. The subject that gets the clearance of a low level can access objects at that level as well as to all objects in lower levels but can't access all objects at higher levels. For example, a subject with a Top-Secret clearance can access Top Secret data, Secret data, and confidential data.

Compartmentalized Environment

In a compartmentalized environment, one security domain is deserted from another. Each domain resides in a separate isolated compartment. The subject must have specific clearance for a specific security domain for accessing an object in that domain.

Hybrid Environment

A hybrid environment has both combinations of hierarchical and compartmentalized concepts. This enables each hierarchical level to contain numerous subdivisions isolated from the rest of the security domain. A specific correct clearance and knowledge of data within a specific compartment are needed to be known by the subject to access the compartmentalized object. Granular control is provided by a hybrid MAC environment over access, which becomes gradually difficult to manage with growth.

Assignment Questions

1. What is an access control method?
2. Mention the advantages of discretionary access controls (DACs)
3. Define role-based access controls
4. Write down the specific characteristic of rule-BAC models
5. Differentiate TBAC with role-BAC

Multiple Choice Questions

1. The task-based access control (TBAC) is quite similar to role-BAC

   (A) True, (B) False

2.   ………………….. is a system that allows the creator, owner, or data custodian of
     an object to define and control access to that object.

     (A) DACs, (B) TBAC, (C) RBAC

3.   ………………. helps for prevention of privilege creep by enforcing the principle
     of least privilege

     (A) DAC, (B) ACL, (C) RBAC, (D) None of these

4.   Who can easily grant multiple permissions simply by adding a new user into
     the appropriate role in role-based access controls?

     (A) Administrators, (B) User, (C) Sender

5.   Which environment contains various classification labels from low security to
     medium security to high security

     (A) Compartmentalized Environment

     (B) Hybrid Environment

     (C) Hierarchical Environment

Answer

1.   True
2.   DACs
3.   RBAC
4.   Administrators
5.   Hierarchical Environment

Summary Questions

1.   What is Mandatory Access Control?
2.   Define software-defined wide area network
3.   What is a Firewall?
4.   What are military security labels?
5.   What is the relationship between each classification or level label in the
     structure?

Chapter 13
Managing the Identity and Access Provisioning Life Cycle

The identity and access provisioning life cycle indicate the formation, management, and deletion of accounts. Provisioning accounts confirm the achievement of appropriate privileges according to task requirements. Allocation of proper privileges according to assigned tasks is confirmed by periodic reviews. Disabling accounts as soon as an employee leaves the company is called revocation and deleting accounts when they are out of necessity.

Although these activities may seem tedious, they are essential to control a system's access authorization and management. The establishment of accurate identity, authentication, authorization, or tracking accountability is not possible without properly defined and maintained user accounts. As stated previously, identification occurs according to the claims of a subjects' identity. This identity is mostly a user account, also including service and computer accounts.

Access control administration is a bundle of tasks and duties involving the management of accounts, access, and accountability during the accounts' life span. These tasks are divided into three main duties of the identity and access provisioning life cycle: provisioning, account review, and account revocation.

Provisioning

The primary step in identity management is creating new accounts and provisioning them by granting them appropriate access to resources. In role-based access control (or group-based), management is often simplified by placing accounts into groups with defined privileges.

The coordination of the formation of user accounts, email authorizations, and other tasks such as the association of physical resources to new users is called provisioning. Provisioning systems and identity management should include a workflow component to maintain industry standards. Workflow allows administrators to specify a sequence of events to add users based on the users' roles and the approval of others in the organization1. An automatic provisioning process controlled by the software is used in some organizations. For example, when an employee is hired, the employee's information is entered into an internal website application by the human resources

© The Author(s), under exclusive license to Springer Nature Singapore Pte Ltd. 2023   173
B. S. Rawal et al., *Cybersecurity and Identity Access Management*,
https://doi.org/10.1007/978-981-19-2658-7_13

department. This application is tied to a database and can automatically create the account and add it to the appropriate groups based on where the new employee will work.2 To keep and maintain account and information, the provisioning process and other identity management operations should be the same for all personnel. Although the way and extent of provisioning customers and partners will differ from employees, requirements of different systems and administration methods are not applicable for different types of users1. When additional privileges are needed, provisioning also occurs during the lifetime of an account. For example, an IT guy assigned to the IT department needs privileges associated with IT. If this person transfers to the sales department, the account needs modification, adding privileges needed to deal with the salesperson.

Password management is another important step in provisioning. Even authorities of small and midsized organizations use multiple passwords for their personal, departmental, and enterprise implementations. To avoid problems like personal creep and excessive privilege, passwords must be checked and altered regularly for compliance regulation and security maintenance. Excessive privilege happens when users tend to acquire more privileges than their assigned work tasks require. On the other hand, creeping privileges involve the illegal accumulation of privileges from a user account even after their job roles have changed or been revoked. Adding or removing users from their roles or groups based on their current job position and assigned work can prevent permission creep and excessive misuse of privilege.

Although password policies and account lockout policies are often considered to be the weakest type of authentication, these are still part of provisioning. But the situation can change through the selection of a strong password and regular alteration. They commonly include the following elements2:

Password length: It has been advised recently to use lengthy passwords with more diversified characters as the tools used to crack passwords are becoming better, and processor strength is increasing.

Complexity: To give it a taste of complexity, passwords should have at least three of the four-character types, which are uppercase, lowercase, numbers, and symbols).

History: Users must avoid reusing the same one as password history will often remember the last 12 or 24 passwords used by an account.

Maximum age: Users should regularly alter passwords to sustain security measurements. Regular users might be required to change their passwords every 45, 60, or 90 days, whereas privileged accounts should do it more often, around every 30 days, and.

Minimum age: There is a time gap between the resetting password, and it is often set to one day. It prevents users from reusing the same one and repetition of resetting their password to bypass the history requirement.

When incorrect passwords attempts are made too many times, account lockout policies take cover of the accounts. For example, access into the account will be held for about 30 min after five failed attempts of password input. The account can be set to remain locked until an administrator unlocks it with verification. Some policies implement a silent notification to let know the authorized user about the failed attempts and intentionally delay login attacks to prevent brute force attacks.

Keep tracking passwords creates problems like forgetting passwords and ending up at the help desk, which eventually increases the cost of management. To solve this problem, password reset systems can be used, enabling users to reset their passwords leaving administrative intervention. A registration process requires users to answer secret questions during resetting the password, which eventually risks the user's identity even before problems are resolved. Previously, hackers have used social engineering methods to collect these secrets and pretend to the user during the reset process. But password reset systems that work through email or personal number verification are merely susceptible to these types of attacks.

To minimize the burden on users to remember passwords, two general approaches have been used: SSO and password synchronization. Password synchronization systems correlate all user passwords to the same word. Doing so helps the user to maintain multiple passwords, which are a bit costly. But there is a more adverse sight of this system for which most of the time it's not recommended. The reason is someone will have all the passwords if he discovers the password to any one of those systems. Although password synchronization is an option for password management, this method is not recommended.

SSO servers are more complex as they work directly with web-based applications intercepting HTTP traffic and responding to password requests. In any system where users access, it stores individual passwords for each system. The SSO server intercepts the request and responds on behalf of the user whenever an application challenges a user for credentials. To implement SSO, legacy applications distinctively require specialized, sometimes custom-made code.

Provisioning in an Organization

Although creating a new user account is usually an easy process, organizational security policy procedures must be used to make it protected and secured. Users shouldn't create their accounts in response to random requests or at an administrator's whim. Rather, proper provisioning ensures that personnel follows specific procedures when creating accounts.

Enrollment or registration is known to be the process of creating a new user account. This process creates a new identity and establishes the factors the system needs to perform for authentication. The first thing an organization needs to ensure is that this process is completed fully and accurately. Also, the organization needs to verify the identity of the individual being enrolled be proved through the means it feels necessary and sufficient. Photo ID, birth certificate, background check, credit check, security clearance verification, FBI database search, and even calling references are all valid forms of verifying a person's identity before enrolling them in any secured system.

Nowadays, many organizations are using automated provisioning systems. For example, after hiring a new employee, the HR department completes initial identification and in-processing steps. Then they forward a request to the IT department to create an account. Users within the IT department enter information about the employee, such as the employee's name and their assigned department, in an application. That application has predefined rules for account creation and uses the rules to create the account. These predefined rules ensure that the system creates accounts consistently. If the policy dictates that username include first and last names, then the application will create a username as patrickjay for a user named Patrick Jay. If an employee with the same name gets hired later, then the second username might be patrickjay2. If the organization is using groups (or roles), the application can automatically add the new user account to the appropriate groups based on the user's department or job responsibilities. The groups will already have appropriate privileges assigned, so this step provisions the account with appropriate privileges.

New employees should receive training on organization security policies and procedures as part of the hiring process. Also, they are usually required to review and sign an agreement committing to uphold the organization's security standards before the hiring process is completed. The agreement often contains an acceptable usage policy.

A user account goes through maintenance throughout its life. Organizations with static organizational hierarchies and low employee turnover or promotion will conduct significantly less account administration than an organization with a flexible or dynamic organizational hierarchy and high employee turnover and promotion rates. Altering rights and privileges are the two things most account maintenance has to deal with. To control how access is changed throughout the life of a user account, procedures similar to those should be used when creating new accounts. Because unauthorized increases or decreases in an account's access capabilities can cause serious security repercussions.

Identity Management

The significant events that occur in an information system need auditing for effective security management and compliance procedures. As the central role of access control is identity management, it should have reliable, tamper-proof logging of all operations such as authenticating with an application, querying database records, and modifying customer information. Although most OSs offer logging functionality, these log files can easily be tampered with. That is why secure logging is a requirement to provide creditable proof during security and regulatory audits. Audit reporting allows managers and administrators to monitor activities and to demonstrate compliance with relevant regulations. To reduce the cost associated with proving compliance, effective audit reporting is crucial.

The traditional organizational boundaries that have marked the reaches of enterprise applications no longer limit the reach of these systems. To efficiently and effectively use multi-organizations systems, commercial and government organizations need to share identity information. For example, engineers at an automobile manufacturer need access to product specifications from a parts supplier. With the thousands of parts required for components and subcomponents, online access to product information is required to support collaborative development. In identity management, provisioning, auditing, and access control are related. Creating an account, which is a provisioning task, should be logged in an audit trail.

Access control details should be included in audit trails as well. Enforcement systems should log attempted violations of access control rules. Even though enforcement systems will vary by platform, a logically centralized view of enforcement audits should be available to systems administrators.

Workflow identity management operations that touch multiple systems should be executed through a workflow mechanism. With workflow, organizations can ensure that proper approvals are always obtained while maintaining business controls. Auditing these approvals enables workflow to provide accountability as it provides information about when the approval happened and by whom. Also, workflows help to enable modular designs of identity management systems. Although identity management operations can indeed be tightly integrated, organizations usually do not deploy a full identity management system at one time. A better approach is to adopt a single module at a time and integrate the modules in a phased implementation. However, since IAM solutions are closely related and require multiple integration points, leading analysts have indicated that the majority of enterprises will be turning to integrated suites from a single vendor rather than struggling with "after the fact" integration, which has been a key cause of failure for many identity projects. Thus, once a loner-view IAM plan is in place, organizations should implement modules that will address the most pressing pain points in the organizations. For example:

- If help desk support costs are increasing rapidly and users are becoming increasingly frustrated with the proliferation of passwords, they must manage, then start with SSO or self-service password reset.
- If your organization has high turnover and has felt the effects of poor account management (e.g., a disgruntled former employee has hacked into one of your servers), then a provisioning system would be a good start.
- If your users need to access a large number of systems and your administrative staff is spending too much time resetting passwords and monitoring access on multiple systems, then look to an enforcement module to start.

Separately deployed modules still need integration—a standardization argument based on widely recognized protocols, like SPML and SAML, and broader initiatives, like the Liberty Alliance. The modules should also maintain tamper-proof audit data to keep pace with compliance with many regulations as integration requires more than workflow to move data.

End-User and Administration Support

Identity management systems ideally provide a single point of access for end-users to manage their identity information and a management console for administrators. Users should have a single resource for common operations such as:

- Updating passwords
- Requesting access to applications and other resources
- Updating personal information in their identity management profiles
- Retrieving security policies.

Administrator management consoles can encompass common operations, such as reviewing alerts from security systems and modifying policies.

Account Review

A periodical review of accounts is required to ensure that security and company policies are implemented and carried out properly. Accounts with top privileges should be checked periodically to detect any anomaly going through the accounts as they are more susceptible to be hacked. Also, task privileges should be granted to a group rather than to an individual. These groups should also be reviewed periodically to prevent excessive privilege and privilege creeping problems. Groups and roles are not the only responsibilities to be monitored. Also, inactive accounts should be checked and disabled periodically to avoid excessive privileges to employees. Nowadays, scripts are being used by many administrators to detect and disable inactive accounts. For example, a script locates accounts which aren't logged into for the past 30 days and disables them automatically. In the same way, group membership of privileged groups can be checked by scripts to remove excess privilege and remove unauthorized accounts. Auditing procedures most often carry out an account review.

Account Revocation

Revocation of account is a concept about which most people know but don't strictly follow. It indicates disabling user accounts of employees when they leave an organization for any reason or take a leave of absence. It's very important for people who are posted in administrative positions. HR departments know when to act as they keep track of employees leaving or changing job positions. They have the full authority to revoke and control access of the leaving employees as soon as possible. They can perform this task because they are aware of when employees are leaving for any reason. The terminated employee can retain access to malicious action. Even if he doesn't, others can use that account for blackmailing, if by any chance, they come across the password.

Instead of being handled by a person, logs can record the activity of the account in the name of the terminated employee. Sometimes the account of a terminated employee can be needed for accessing encrypted data, so it can't be deleted right away. Authorities can delete it when that is no longer needed. Terminated accounts can be deleted within 30 days but vary depending on the needs of the organization according to ongoing administrative oversight (Fig. 13.1).

Assignment Questions

1. Define provisioning
2. What is the role of password management in provisioning?
3. Give notes on password policies and account lockout policies
4. Address the importance of identity management
5. Describe the process of provisioning, auditing, and access control in identity management.

Fig. 13.1   Identity management and access provisioning life cycle

Multiple Choice Questions

1. .... accounts confirm the achievement of appropriate privileges according to task requirements.
   (A) Provisioning (B) Disabling
2. Which management is often simplified by placing accounts into groups with defined privileges?
   (A) DAC (B) RBAC
3. .... is known to be the process of creating a new user account
   (A) Enrollment (B) Registration (C) A and B
4. Which management provides a single point of access for end-users to manage their identity information and a management console for administrators?
   (A) Identity (B) User (C) A and B
5. Who should have a single resource for common operations?
   (A) Users (B) Admin (C) Enrollment

Answers

1. Provisioning
2. RBAC
3. A and B
4. Identity
5. Users

Summary Questions

1. What is SSO?
2. What is HTTP traffic?
3. What is the IAM plan?
4. Differentiate SPML and SAML?
5. What is account revocation?

Chapter 14
Conduct Security Control Testing

Design and Validate Assessment, Test, and Audit Strategies

Security assessment and testing programs are the two main tasks of an information security team. It includes regularly verifying that an organization's security controls are adequate and that those security controls are functioning correctly and effectively safeguarding information assets with the help of tests, assessments, and audits. In this section, you will learn about the three major components of a security assessment program:

- Security tests
- Security assessments
- Security audits.

Security Testing

Security tests verify if control is functioning properly. These tests include tool-assisted penetration tests, automated scans, and manual attempts to undermine security. Security testing should be performed regularly, and while performing it, each of the vital security controls protecting an organization should be checked properly.

When scheduling security controls for review, information security managers should consider the following factors:

- Availability of security testing resources
- The criticality of the systems and applications protected by the tested controls
- Impact of the test on normal business operations
- The sensitivity of the information contained on tested systems and applications
- Rate of change of the control configuration
- Other changes in the technical environment that may affect the control performance
- Likelihood of a technical failure of the mechanism implementing the control
- Difficulty and time required to perform a control test
- Likelihood of a misconfiguration of the control that would jeopardize the security

© The Author(s), under exclusive license to Springer Nature Singapore Pte Ltd. 2023 181
B. S. Rawal et al., *Cybersecurity and Identity Access Management*,
https://doi.org/10.1007/978-981-19-2658-7_14

- A risk that the system will come under attack.

Security teams to design and validate a comprehensive assessment and testing strategy after assessing each of these factors. The process may include frequent use of automated testing along with performing manual testing from time to time. For example, a credit card processing system may undergo automatic vulnerability scanning regularly with immediate alerts to administrators when the scan detects a new vulnerability. Automated scans are easy to run as they don't require any work from the administrators once they are configured. The security team may hire an external consultant to perform a manual penetration test to complement those automated scans. This type of manual operation usually costs a significant amount of money. That's why usually, these tests are performed on an annual basis to minimize costs and disruption to the business.

Of course, it's not sufficient to perform security tests. The results of those tests must also be carefully reviewed to ensure that each test was successful. In some cases, these reviews consist of manually reading the test output and verifying that it is completed successfully. Some tests require human interpretation and must be performed by trained analysts. Other reviews may be automated, performed by security testing tools that verify the successful completion of a test, log the results, and remain silent unless there is a significant finding. Once the system detects an issue that requires the attention of an admin, it triggers an alert, sends an email or text message, or automatically opens a trouble ticket, depending on how severe the issue is and the administrator's preference.

Security Assessments

Security assessments are comprehensive reviews of the security of a system, application, or other tested environments. To identify the vulnerabilities in a tested environment that may compromise, a trained information security professional performs a risk assessment as part of the security assessment. Then based on his findings, he makes recommendations for remediation. The use of security testing tools is usually included in security assessments. However, it goes beyond manual penetration tests and automated scanning. They also have a thoughtful review of the threat environment, current and future risks, and the value of the targeted environment.

An assessment report addressed to management is usually the primary work product of a security assessment. It contains the results of the evaluation in nontechnical language and concludes with specific recommendations for improving the security of the tested environment.

Security Audits

Security audits and security assessments both use similar techniques. However, security audits must be performed by independent auditors, which isn't a criterion for security assessments. While an organization's security staff may routinely perform security tests and assessments, this is not the case for audits. Assessment and testing results are meant for internal use only and are designed to evaluate controls to find potential improvements. On the other hand, audits are evaluations performed to demonstrate the effectiveness of controls to a third party. It will be a conflict of interest if the team who design, implement, and monitor controls for an organization evaluate the effectiveness of those controls. Auditors provide an impartial, unbiased view of the state of security controls. They write reports similar to security assessment reports, but those reports are intended for different audiences that may include an organization's board of directors, government regulators, and other third parties.

There are two main types of audits: internal audits and external audits.

An organization's internal audit staff performs internal audits. They are typically intended for internal audiences. The reporting line of the internal audit staff performing these audits that are entirely independent of the functions they evaluate. In many organizations, the chief audit executive reports directly to the president, chief executive officer, or similar roles. The chief audit executive may also have reporting responsibility now to the organization's governing board.

Outside auditing firms perform external audits. As the auditors performing these audits theoretically have no conflict of interest with the organization itself, these assessments have a high degree of external validity. There are thousands of firms that perform external audits, but most people place the highest credibility with the so-called Big Four audit firms:

- Ernst and Young
- Deloitte and Touche
- PricewaterhouseCoopers
- KPMG.

Audits performed by these firms are generally considered acceptable by most investors and governing body members.

Information security professionals are often asked to participate in both internal and external audits. Through interviews and written documentation, they provide information about security controls to auditors. The auditors may also request the participation of security staff members in the execution of control evaluations. Auditors are provided with access to all information within an organization, and security staff should comply with those requests, consulting with management as needed.

Performing Vulnerability Assessments

Vulnerability assessments are some of the most critical testing tools that information security professionals have to use regularly. Security professionals get a perspective on the weaknesses in an application's technical controls or a system by utilizing vulnerability scans and penetration tests.

Vulnerability Scans

Vulnerability scans automatically investigate applications, systems, and networks, looking for weaknesses that may be exploited by an attacker. The scanning tools used for these tests provide quick, point-and-click tests. With most tools, administrators can schedule scanning repeatedly and get reports that show differences between scans performed on different days. Administrators can use this data to figure out the changes in their security risk environment.

There are three main vulnerability scans: network discovery scans, network vulnerability scans, and web application vulnerability scans. A wide variety of tools perform each of these types of scans.

Network Discovery Scanning

To search for systems with open network ports, network discovery scanning uses various techniques to scan a range of IP addresses. Checking systems for vulnerabilities isn't the application of network discovery scanners. Instead, they are used to generate a report showing the methods detected on a network and the list of ports exposed through the network and server firewalls that lie on the network path between the scanner and the scanned system.

Network discovery scanners use many different techniques to identify open ports on remote systems. The more common methods are as follows:

- TCP SYN scanning sends a single packet to each scanned port with the SYN flag set. It indicates a request to open a new connection. If the scanner receives a response with the SYN and ACK flags set, this means that the system is moving to the second phase in the three-way TCP handshake and that the port is open. TCP SYN scanning is also known as "half-open" scanning.
- TCP Connect scanning opens a full connection to the remote system on the specified port. This scan type is used when the user running the scan does not have the necessary permissions to run a half-open scan.
- TCP ACK scanning sends a packet with the ACK flag set, indicating that it is part of an open connection.
- Xmas scanning sends a packet with the FIN, PSH, and URG flags set. A packet with so many flags set is said to be "lit up like a Christmas tree," leading to the scan's name.

An open-source tool called Nmap is the most widely used tool for network discovery scanning. It was initially released in 1997. But remarkably, it is still being maintained and used today.

Network Vulnerability Scanning

Network vulnerability scans go further than discovery scans. After detecting open ports, they don't stop their examination process but continue to look for vulnerabilities on a targeted system or network. Also, they perform tests to identify whether a system is susceptible to each vulnerability in the system's database.

A scanner uses the tests in its database to test a system for vulnerabilities. In some cases, the scanner may not have enough information to determine that a vulnerability exists conclusively, and it reports a vulnerability when there is no problem. It is called a false positive report and is sometimes seen as a nuisance to system administrators. The incident that is more dangerous than a false positive is a false negative report. A false-negative report is known to be when the vulnerability scanner misses a vulnerability and fails to alert the administrator to the presence of a dangerous situation.

By default, unauthenticated scans are run by network vulnerability scanners. When they test target systems, they are not provided with passwords or any other special information to give the special scanner access. This allows the scanner to perform the scanning process from an attacker's perspective. The downside of it is that it limits the ability of the scanner to evaluate possible vulnerabilities fully.

Performing authenticated scans of systems is one way to improve the accuracy of the scanning and reduce false positive and false negative reports. In this approach, the scanner has read-only access to the servers being scanned. It can use this access to read configuration information from the target system and use it when analyzing vulnerability testing results.

Web Vulnerability Scanning

Web applications pose a significant risk to enterprise security. The servers that run many web applications must expose their services to internet users. Firewalls and other security devices typically contain rules allowing web traffic to pass through to web servers unfettered. The applications running on web servers are complex and often have privileged access to underlying databases. Attackers often use SQL injection and other attacks to exploit these circumstances and target the flaws in the security design of web applications.

Web vulnerability scanners are special-purpose tools that test web applications for known vulnerabilities. As they can discover flaws that are not visible to network vulnerability scanners, they plan a vital role in any security testing program. When an administrator runs a web application scan, the tool tests the web application using automated techniques that manipulate inputs and other parameters to identify web vulnerabilities. Finally, a report is provided by the tool that contains its findings and recommendation for vulnerability remediation techniques.

The operation of web vulnerability scans and network vulnerability scans is quite similar. Both examine services running on a server for known vulnerabilities. The difference is that network vulnerability scans generally don't dive deep into the structure of web applications. In contrast, web application scans don't look at services other than those supporting web services. Although many network vulnerability scanners perform basic web vulnerability scanning tasks, deep-dive web vulnerability scans require specialized, dedicated web vulnerability scanning tools.

Web vulnerability scans are an essential component of an organization's security assessment and testing program. It's a good practice to run scans in the following circumstances:

- When you begin performing web vulnerability scanning for the first time, you should scan all applications. This will detect issues with legacy applications.
- Before moving any new application into a production environment for the first time, perform a proper scan.
- Scan any modified application before the code changes move into production.
- Scan all applications regularly. Limited resources may require scheduling these scans based on the priority of the application. For example, you may wish to scan

web applications that interact with sensitive information more often than those that do not.

Penetration Testing

Penetration testing takes vulnerability testing one step further by actually attempting to exploit systems. As vulnerability scans do generally not take offensive action against the targeted system, they can merely examine the presence of a vulnerability. On the other hand, the main focus of a security professional performing penetration tests is to break into a targeted system or application to defeat security controls to demonstrate the flaw.

When performing a penetration test, the security professional typically targets a single system or set of systems and uses many different techniques to gain access. The process may include the following:

- Performing necessary reconnaissance to determine system function such as visiting websites hosted on the system
- Network discovery scans to identify open ports
- Network vulnerability scans to identify unpatched vulnerabilities
- Web application vulnerability scans to identify web application flaws
- Use of exploit tools to automatically attempt to defeat the system security
- Manual probing and attack attempts.

A commonly used tool called Metasploit is used by penetration testers to execute exploits against targeted systems automatically. Penetration testers may be company employees who perform these tests as part of their duties or external consultants hired to perform penetration tests. The tests are typically categorized into three groups:

White Box Penetration Test

In this test type, the attackers are provided with detailed information about the systems they target. Many of the reconnaissance steps that typically precede attacks are skipped to shorten the time of the attack and increasing the likelihood that it will find security flaws.

Black Box Penetration Test

In this penetration test, the attackers are not provided with any information before the attack. It is done to simulate an external attacker trying to gain access to information about the business and technical environment before engaging in an attack.

Gray Box Penetration Test

They are also known as partial knowledge tests and are usually chosen to balance the advantages and disadvantages of white and black box penetration tests. This is particularly common when black box results are desired, but costs or time constraints mean that some knowledge is needed to complete the testing.

Penetration tests are time-consuming and require specialized resources, but they play an essential role in the ongoing operation of a sound information security testing program.

Testing Your Software

In system security, the software is considered to be one of the most critical components. Think about the following characteristics common to many applications in use throughout the modern enterprise:

- Often privileged access to the operating system, hardware, and other resources is given to software applications.
- Software applications routinely handle sensitive information, including credit card numbers, social security numbers, and proprietary business information.
- Many software applications rely on databases that also contain sensitive information.
- The software performs functions that are critical to businesses. That's why it is considered to be the heart of modern enterprise. Software failures can disrupt businesses with severe consequences.

These are only a few of many reasons that make careful testing of software crucial to the confidentiality, integrity, and availability requirements of every modern organization. Now, we will be discussing some of the types of software testing that you may integrate into your organization's software development life cycle.

Code Review and Testing

Performing code review and testing is one of the most critical components of a software testing program. These procedures provide third-party reviews of the work performed by developers before moving code into a production environment. Code reviews and tests may discover security, performance, or reliability flaws in applications before they go the life and negatively impact business operations.

Code Review

The foundation of software assessment programs is code reviews. It is also known as a "peer review." During this process, developers other than the one who wrote the code review it to find defects. If no flaw in the code is found, the application is approved to move into a production environment. However, if deficiencies in the code are found, the code is sent back to the original developer with recommendations for rework of issues detected during the review.

Code review takes many different forms and varies in formality from organization to organization. Fagan inspections, which is known to be one of the most formal code review processes, follow a rigorous review and testing process with six steps:

1. Planning
2. Overview
3. Preparation
4. Inspection

5.   Rework
6.   Follow-up.

An overview of the Fagan inspection appears in Fig. 14.1. Each of these steps has well-defined entry and exit criteria that must be met before the process may formally transition from one stage to the next.

In highly restrictive environments where code flaws may have a catastrophic impact, the Fagan inspection level of formality is usually found. Most organizations use less rigorous processes using code peer review measures that include the following:

- Developers walking through their code in a meeting with one or more other team members
- A senior developer performing manual code review and signing off on all code before moving to production
- Use of automated review tools to detect common application flaws before moving to production.

Based on business requirements and software development culture, each organization should select a code review process.

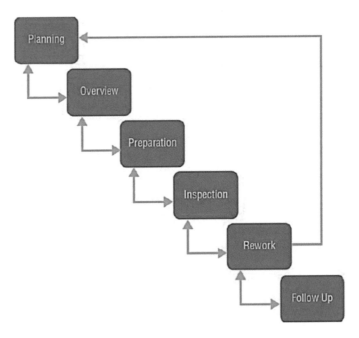

Fig. 14.1 Fagan inspection process

Static Testing

Static testing is used to test the security of software without running it. For this, either the source code or the compiled application is analyzed. Static analysis usually involves using automated tools designed to detect common software flaws, such as buffer overflows. Application developers in mature development environments use static analysis tools throughout the design, build, and test process.

Dynamic Testing

In dynamic testing, the security evaluation of the software is done in a runtime environment and is often the only option for organizations deploying applications written by someone else.

In those cases, testers often do not have access to the underlying source code. Using web application scanning tools to detect cross-site scripting, SQL injection, or other flaws in web applications is a typical example of dynamic software testing. To avoid any unwanted interruption of service, you need to be careful while performing dynamic tests in a production environment.

Dynamic testing may include the use of synthetic transactions to verify system performance. Synthetic transactions are scripted dealings with known expected results. The synthetic transactions are run by the testers against the tested code and then compare the output of the transactions to the expected state. The code is considered to have flaws if any deviation between the actual and expected results is present. In cases like this, further investigation is done on the code.

Fuzz Testing

Fuzz testing is a specialized dynamic testing technique that provides many different types of input to software to stress its limits and find previously undetected flaws. In fuzz testing, the software is supplied with invalid input, either specially crafted to trigger known software or randomly generated. The fuzz tester then monitors the performance of the application, watching for software crashes, buffer overflows, or other undesirable and/or unpredictable outcomes.

There are two main categories of fuzz testing:

Mutation (Dumb) Fuzzing takes previous input values from the actual operation of the software and manipulates (or mutates) it to create fuzzed input. It might alter the characters of the content, append strings to the end of the content, or perform other data manipulation techniques.

Generational (Intelligent) Fuzzing develops data models and creates new fuzzed input based on an understanding of the types of data used by the program.

Interface Testing

Interface testing is an essential part of the development of complex software systems. Usually, when developing complex software, multiple teams of developers work on different aspects of it. However, to meet the business objectives, the other parts must function together. The handoffs between these separately developed modules use well-defined interfaces so that the teams may work independently. To ensure that the modules will work together correctly when all of the development efforts are complete, interface testing assesses them against the interface specifications.

Three types of interfaces should be tested during the software testing process:

Application programming interfaces (APIs) offer a standardized way for code modules to interact and be exposed to the outside world through web services. Developers must test APIs to ensure that they enforce all security requirements.

User interfaces (UIs) examples include graphical user interfaces (GUIs) and command-line interfaces. UIs provide end-users with the ability to interact with the software. Interface tests should consist of reviews of all user interfaces to verify that they function correctly.

Physical interfaces exist in some applications that manipulate machinery, logic controllers, or other objects in the physical world. Software testers should pay careful attention to physical interfaces because of the potential consequences if they fail.

Interfaces provide essential mechanisms for the planned or future interconnection of complex systems. To facilitate interactions between different software packages, the Web 2.0 world depends on the availability of these interfaces.

Misuse Case Testing

In some software, there some ways that users can try to misuse the application. For example, in banking software, a user can gain access to another user's account by manipulating input strings. They might also try to withdraw funds from a charge that is already overdrawn. To evaluate the vulnerability of software to these known risks, software testers use a process known as misuse case testing or abuse case testing.

Testers first identify the known misuse cases in misuse case testing. Then they try to exploit those use cases with automated and/or manual attack techniques.

Assignment Questions

1. What are the three major components of a security assessment program? Explain
2. What are the factors information security managers should consider when scheduling security controls for review?
3. Explain security audit in detail.
4. What is web vulnerability scanning?
5. Explain the process includes in a penetration test.

Multiple Choice Questions

1. Security assessment and testing programs are the two main tasks of an information security team, true or false
      (A) True (B) false
2. How many major components are there for a security assessment program?
      (A) 1 (B) 2 (C) 3
3. ....must be performed by independent auditors
      (A) security audits (B) security assessments (C) none of the above
4. An organization's internal audit staff performs internal audits. True or false
      (A) True (B) false

5.   Which assessments have a high degree of external validity?
     (A) Internal audits (B) security assessments (C) external audit

Answers

1.   True
2.   3
3.   Security audits
4.   True
5.   External audit.

Summary Questions

1.   What are the factors information security managers should consider when scheduling security controls for review?
2.   Explain the methods to identify open ports on remote systems
3.   What is network vulnerability scanning? Explain
4.   What is the process including in the penetration test?
5.   Explain the Fagan inspection process.

Chapter 15
Collect Security Process Data

Organizations must collect the appropriate security process data once security controls are tested. NIST SP 800–137 provides guidelines for developing an information security continuous monitoring (ISCM) program. Security process data should include account management, management review, key performance and risk indicators, backup verification data, training and awareness, and disaster recovery and business continuity. It's the job of the security professionals to ensure that the security process data have those included.

NIST SP 800–137

According to NIST SP 800-137, ISCM is defined as maintaining ongoing awareness of information security, vulnerabilities, and threats to support organizational risk management decisions.

Organizations should take the following steps to establish, implement, and maintain ISCM:

1. Define an ISCM strategy based on risk tolerance that maintains clear visibility into assets, awareness of vulnerabilities, up-to-date threat information, and mission/business impacts.
2. Establish an ISCM program that includes metrics, status monitoring frequencies, control assessment frequencies, and an ISCM technical architecture.
3. Implement an ISCM program and collect the security-related information required for metrics, assessments, and reporting. Automate the collection, analysis, and reporting of data where possible.
4. Analyze the data collected, report findings, and determine the appropriate responses. It may be necessary to collect additional information to clarify or supplement existing monitoring data.
5. Respond to findings with technical, management, and operational mitigating activities or acceptance, transference/sharing, or avoidance/rejection.
6. Review and update the monitoring program, adjusting the ISCM strategy and maturing measurement capabilities to increase visibility into assets and

© The Author(s), under exclusive license to Springer Nature Singapore Pte Ltd. 2023   193
B. S. Rawal et al., *Cybersecurity and Identity Access Management*,
https://doi.org/10.1007/978-981-19-2658-7_15

awareness of vulnerabilities, further enable data-driven control of the security of an organization's information infrastructure, and increase organizational resilience.

Account Management

The addition and deletion of accounts that are granted access to systems or networks are parts of account management. Also, changing the permissions or privileges granted to those accounts is involved in account management. If account management is not monitored and recorded properly, organizations may discover that accounts have been created for the sole purpose of carrying out fraudulent or malicious activities. In account management, two-person controls should be used, often involving one administrator who creates accounts and another who assigns those accounts the appropriate permissions or privileges.

Escalation and revocation are two terms that are important to security professionals. When a user account is granted more permission based on new job duties or a complete job change, account escalation occurs. Before changing the current permissions or privileges of a user, security professionals should fully analyze that user's needs, making sure to grant only permissions or privileges that are needed for the new task and to remove those that are no longer needed. Because without such proper analysis, users may be able to retain permissions that cause possible security issues because the separation of duties is no longer retained. For example, suppose a user is hired in the accounts payable department to print out all vendor checks. Later this user receives a promotion to approve payment for the same accounts. If this user's old permission to print checks is not removed, this single user would be able to both approve the checks and print them, which is a direct violation of the separation of duties.

Account revocation occurs when a user account is revoked because a user is no longer with an organization. Security professionals must keep in mind that there will be objects that belong to this user. If the user account is simply deleted, access to the objects owned by the user may be lost. It may be a better plan to disable the account for a certain period. Account revocation policies should also distinguish between revoking an account for a user who resigns from an organization and revoking an account for a user who is terminated.

Management Review and Approval

Management review of security process data should be mandatory. An organization might be collecting a huge amount of data on its security processes. But unless it is reviewed by an administrator, the data is useless. That's why organizations should establish guidelines and procedures to ensure that management review occurs on time. Without regular review, even the most minor security issue can be quickly turned into a major security breach.

Management review should include an approval process. The security professionals will give recommendations to the management, and the management will review them. Based on the data provided, the management will either approve or reject the recommendations. If alternatives are given, management should approve the alternative that best satisfies the organizational needs. Security professionals should ensure that the reports provided to management are as comprehensive as possible so that all the data can be analyzed to ensure the most appropriate solution is selected.

Key Performance and Risk Indicators

Organizers can identify when security risks are likely to occur by using key performance and risk indicators of security process data. Key performance indicators (KPIs) enable organizations to determine whether levels of performance are below or above established norms. Whereas key risk indicators (KRIs) allow organizations to identify whether certain risks are more or less likely to occur.

For improving Critical Infrastructure Cybersecurity, the NIST has released a Framework which is known as the Cybersecurity Framework. The Cybersecurity Framework focuses on using business drivers to guide cybersecurity activities and considering cybersecurity risks as part of the organization's risk management processes. The framework consists of three parts: The Framework Core, the Framework Profiles, and the Framework Implementation Tiers.

The Framework Core is a set of cybersecurity activities, outcomes, and informative references that are common across critical infrastructure sectors, providing detailed guidance for developing individual, organizational profiles. The Framework Core consists of five concurrent and continuous functions: identity, protect, detect, respond, and recover.

After each function is identified, categories and subcategories for each function are recorded. The Framework Profiles are developed based on the business needs of the categories and subcategories. Through the use of the Framework Profiles, the framework helps an organization align its cybersecurity activities with its business requirements, risk tolerances, and resources.

The Framework Implementation Tiers provide a mechanism for organizations to view and understand the characteristics of their approach to managing cybersecurity risk. The following tiers are used: Tier 1, partial; Tier 2, risk-informed; Tier 3, repeatable; and Tier 4, adaptive.

Organizations will continue to have unique risks—different threats, different vulnerabilities, and different risk tolerances. Therefore, how they implement the practices in the framework will vary. Ultimately, the framework is aimed at reducing and better managing cybersecurity risks and is not a one-size-fits-all approach to managing cybersecurity.

System Resilience and Fault Tolerance

As security process data contains important user information and other technical information, it needs to be stored carefully. When choosing storage for such important data, we need to ensure that it doesn't have a single point of failure and has great fault tolerance and system resilience.

A single point of failure is any component that can cause an entire system to fail. For example, if the data of a computer is stored on a single disk, failure of the disk can cause the computer to fail, so the disk is a single point of failure. If a database-dependent website includes multiple web servers all served by a single database server, the database server is a single point of failure.

Fault tolerance is the ability of a system to continue operating even after suffering a fault. To achieve fault tolerance, redundant components such as additional disks within a redundant array of inexpensive disks (RAID) array or additional servers within a failover clustered configuration is added.

A system's ability to maintain an acceptable level of service during an adverse event is known as system resilience. This could be a hardware fault managed by fault-tolerant components, or it could be an attack managed by other controls such as effective intrusion detection and prevention systems. In some contexts, it refers to the ability of a system to return to a previous state after an adverse event.

Protecting Hard Drives

A redundant array of disks (RAID) array is a common way of adding fault tolerance and system resilience to computers. Two or more disks are included in a RAID array, and most RAID configurations are capable of operating even after one of the disks fails. Some of the common RAID configurations are as follows:

RAID-0: It is also known as striping. It uses two or more disks and is used to improve the disk subsystem performance. However, it does not provide fault tolerance.

RAID-1: It is also called mirroring. It uses two disks, and both of the disks hold the same data. If one disk fails, the other disk includes the data so a system can continue to operate after a single disk fails. Based on which drive fails and the hardware used, the system may be able to continue its operation without any intervention, or the system may need to be manually configured to use the drive that didn't fail.

RAID-5: It is also called striping with parity. It uses three or more disks with the equivalent of one disk holding parity information. If any single disk fails, the RAID array will continue to operate, though it will be slower.

RAID-10: This is also known as RAID 1 + 0 or a stripe of mirrors. It is configured as two or more mirrors (RAID-1) configured in a striped (RAID-0) configuration. At least, four disks are used in it. However, it can support more as long as an even number of disks are added. It will continue to operate even if multiple disks fail, as long as at least one drive in each mirror continues to function. For example, if it had two mirrored sets (called D1 and D2 for his example), it would have four disks in total. If one drive on D1 and one drive on D2 failed, the array would still function properly. But if two drives of D1 or D2 failed at once, the whole array would fail.

Fault tolerance and backup are two separate things. Occasionally, management may balk at the cost of backup tapes and point to the RAID, saying that the data is already backed up. However, if a catastrophic hardware failure destroys a RAID array, all the data is lost unless a backup exists. Similarly, if data is deleted by accident or gets corrupted, it cannot be restored unless a back exists.

RAID solutions can be both software- and hardware-based. In software-based systems, the disks in the array are managed by the operating system. Therefore,

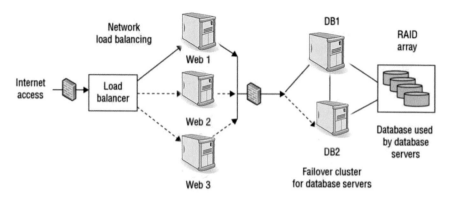

Fig. 15.1 Failover cluster with network load balancing

the overall system performance is affected. However, they are relatively inexpensive since they don't require any additional hardware other than the additional disk(s).

Hardware RAID systems are more expensive. But its reliability and efficiency outweigh the costs when used to increase the availability of a critical component. Spare drives are usually included in hardware-based RAID arrays. These spare drives can be logically added to the array when needed. For example, a hardware-based RAID-5 could include five disks, with three disks in a RAID-5 array and two spare disks. If one disk fails, the hardware senses the failure and logically swaps out the faulty drive with a good spare. Additionally, most hardware-based arrays support hot-swapping, allowing technicians to replace failed disks without powering down the system. A cold swappable RAID requires the system to be powered down to replace a faulty drive.

Protecting Servers

We can add fault tolerance to critical servers with failover clusters. Two or more servers are included in a failover cluster. If something causes one of the servers to fail, another server in the cluster takes over its load. This process happens automatically and is called failover. Failover clusters can include multiple servers, and they can also provide fault tolerance for multiple services or applications.

As an example of a failover cluster, consider Fig. 15.1. It shows multiple components put together to provide reliable web access for a heavily accessed website that uses a database. The two database servers DB1 and DB2 are configured in a failover cluster. Normally, only one of the two servers will function as the active database server, and the second one will be inactive. For example, DB1 will perform all the database services for the website if it is an active server. DB2 will continuously monitor DB1 to ensure it is functioning properly. If DB2 senses a failure in DB1 at any given time, it will cause the cluster to fail over to DB2 automatically.

In Fig. 15.1, you can see that both DB1 and DB2 have access to the data in the database. To provide fault tolerance for the disks, the data is stored on a RAID array. Also, a network load-balancing cluster is used for configuring the three web servers. The function of the load balancer is to balance the client load across the three servers. It can be both hardware- and software-based. It also makes it easy to add additional web servers to handle the increased load while also balancing the load among all the servers. If any of the servers fail, the load balancer can sense the failure and stop sending traffic to that server. Although network load balancing is primarily used to increase the scalability of a system so that it can handle more traffic, it also provides a measure of fault tolerance.

Protecting Power Sources

An uninterruptible power supply (UPS), a generator, or both can be used to add fault tolerance for power sources. UPS is used to get battery-supplied power for a short time between 5 and 30 min. At the same time, a generator is used for a long-term power supply when a power outage occurs. The goal of a UPS is to provide power long enough to complete a logical shutdown of a system or until a generator is powered on and provides stable power.

In real life, commercial power suffers from a wide assortment of problems. A spike is a quick instance of an increase in voltage, whereas a sag is a quick instance of a reduction in voltage. If the power stays high for a long time, it's called a surge rather than a spike. If it remains low for a long time, it's called a brownout. Occasionally, power lines have noise on them called transients that can come from many different sources. All of these issues can cause problems for electrical equipment. A basic UPS, which is also known as offline or standby UPS, provides battery backup and surge protection. It is plugged into commercial power, and critical systems are plugged into the UPS system. During a power outage, the battery backup will provide continuous power to the systems for a short time. Along with battery backup and surge protection, line-interactive UPS offers some additional benefits. They include a variable voltage transformer that can adjust to the overvoltage and undervoltage events without draining the battery.

Generators are used to power systems during long-term power outages. The length of time that a generator will provide power is dependent on the fuel, and a site can stay on generator power as long as it has fuel. Generators commonly use diesel fuel, natural gas, or propane.

Assignment Questions

1. Explain the steps an organization should take to establish, implement, and maintain ISCM.
2. What are RAID configurations? Explain
3. Briefly explain account management
4. Explain NIST SP 800–137
5. What do you mean by fault tolerance? Explain

Multiple Choice Questions

1. Escalation and -------------------- are two terms that are important to security professionals

   (A) Revocation (B) administrator (C) none
2. Account revocation occurs when a user account is revoked because a user is no longer with an organization, true or false
3. Management review should include

   (A) Security (B) Monitoring (C) an approval process
4. ------------------ is a common way of adding fault tolerance and system resilience to computers

   (A) RAID (B) RAFD (C) RFID
5. RAID-0 It is also known as

   (A) Mirroring (B) Striping (C) striping with parity

Answers

1. Revocation
2. True
3. An approval process
4. RAID
5. Striping

Summary Questions

1. Explained RAID configurations.
2. Draw and explain failover cluster with network load balancing.
3. Explain the steps an organization should take to establish, implement, and maintain ISCM.
4. Describe account management.
5. Explain NIST SP 800–137.

Chapter 16
Recovery Strategies for Database

Vital activities of an organization such as process and track operations, logistics, and sales are greatly dependent on its database. This is why database recovery techniques should be a part of an organization's disaster recovery plans. For the same reason, a database specialist who can provide input as to the technical feasibility of various ideas is wise to have on the DRP team. After all, you shouldn't allocate several hours to restore a database backup when it's impossible to complete a restoration in less than half a day!

In the following section, we will be talking about the three main techniques that are used to create offsite copies of database content: electronic vaulting, remote journaling, and remote mirroring. Each of the techniques has its benefits and drawbacks. So, based on your organization's computing requirements and available resources, you need to figure out which one will work best for it.

Electronic Vaulting

In electronic vaulting, database backups are moved to a remote site using bulk transfers. The remote location may be a dedicated alternative recovery site (such as a hot site) or simply an offsite location managed within the company or by a contractor to maintain backup data. If you decide to use electronic vaulting, then you need to keep in mind that there will be a significant delay between a disaster declared and the time your database is ready for operation with current data. For activating a recovery site, appropriate backups from the electronic vault need to be retrieved and applied to the soon-to-be production servers at the recovery site by technicians. To ensure that you don't have any issues with your backup, you need to test your electronic vaulting setup from time to time. Giving your disaster recovery personnel a "surprise test," asking them to restore data from a certain day, can be a great method for this.

Remote Journaling

The process of transferring data with remote journaling is relatively faster. Although data transfers still happen in a bulk transfer mode, they happen more frequently, usually once every hour and sometimes even more frequently. Unlike electronic

© The Author(s), under exclusive license to Springer Nature Singapore Pte Ltd. 2023
B. S. Rawal et al., *Cybersecurity and Identity Access Management*,
https://doi.org/10.1007/978-981-19-2658-7_16

vaulting scenarios, where entire database backup files are transferred, remote journaling setups transfer copies of the database transaction logs containing the transactions that occurred since the previous bulk transfer. Remote journaling and electronic vaulting are similar in a way that transaction logs transferred to the remote site are maintained in a backup device but not applied to a live database server. When a disaster is declared, technicians retrieve the appropriate transaction logs and apply them to the production database.

Remote Mirroring

Remote mirroring is the most advanced database backup solution. That's why it is also the most expensive one. With remote mirroring, a live database server is maintained at the backup site. As the database modifications are being applied to the production server at the primary site, the remote server keeps receiving copies of them simultaneously. That's why the mirrored server stays ready to take over an operational role at a moment's notice. For organizations seeking to implement a hot site, remote mirroring is a popular database backup strategy. However, when considering the application of a remote mirroring solution on your organization's database, be sure to be aware of the infrastructure and personnel costs required to support the mirrored server as well as the processing overhead that will be added to each database transaction on the mirrored server.

Recovery Plan Development

Once you have established your priorities for the business unit and have a good idea of the appropriate alternative recovery sites for your organization, it is time to figure out a true disaster recovery plan. It's not something that you'll be able to do in just one sitting. Most likely, the DRP teams will have to come up with multiple plans and then analyze those plans based on the factors like the operational needs of critical business units and falls within the resource, time, and expense constraints of the disaster recovery budget and available personnel they will decide which one will work the best. Depending on the size of your organization and the number of people involved in the DRP effort, it may be a good idea to maintain multiple types of plan documents intended for different audiences. The following list includes various types of documents worth considering:

- Executive summary providing a high-level overview of the plan
- Department-specific plans
- Technical guides for IT personnel responsible for implementing and maintaining critical backup systems
- Checklists for individuals on the disaster recovery team
- Full copies of the plan for critical disaster recovery team members.

Using custom-tailored documents becomes especially important when a disaster occurs or is imminent.

Personnel who need to refresh themselves on the disaster recovery procedures that affect various parts of the organization will be able to refer to their department-specific plans. It will help critical disaster recovery team members to have checklists to help guide their actions amid the chaotic atmosphere of a disaster. IT personnel will have technical guides helping them get the alternate sites up and running. Finally, managers and public relations personnel will have a simple document that walks them through a high-level view of the coordinated symphony that is an active disaster recovery effort without requiring interpretation from team members busy with tasks directly related to that effort.

Emergency Response

As soon as the essential personnel recognizes that a disaster is in progress or is imminent, they need to start following their disaster recovery plan. To ensure that they can do it without any issues, the plan should contain simple yet comprehensive instructions. Depending on the type of personnel responding to the incident, the nature of the disaster, and the time available before facilities need to be evacuated and/or equipment shut down, and these instructions will vary. For example, instructions for a large-scale fire will be much more concise than the instructions for how to prepare for a hurricane that is still 24 h away from a predicted landfall near an operational site. Emergency response plans are often put together in the form of checklists provided to responders. When creating such checklists, you need to ensure that the checklists are arranged in order of priority, with the most important task first! It's most likely that amid a crisis, the responders will not be able to complete the entire checklist. For this reason, essential tasks (such as "Activate the building alarm") should be placed on top of the checklist. The lower an item on the list, the lower the likelihood that it will be completed before an evacuation/shutdown takes place.

Personnel and Communications

A list of personnel to contact in the event of a disaster should also be a part of a disaster recovery plan. Usually, this includes key members of the DRP team as well as the personnel who execute critical disaster recovery tasks throughout the organization. This response checklist should contain multiple means of contact, such as mobile phone numbers and pager numbers. Also, there should be backup contacts for each role so that if the primary contact can't reach the recovery site for some reason, the backup will perform the duty.

Assessment

The first task of a disaster recovery team after arriving on-site is to assess the situation. This assessment usually occurs in a few steps. It starts with some responders performing a very simple assessment to triage activity and get the disaster response underway. As the incident progresses, more detailed assessments will take place to gauge the effectiveness of disaster recovery efforts and prioritize the assignment of resources.

Backups and Offsite Storage

The backup strategy of your organization should be fully addressed by the disaster recovery plan, especially the technical guide. This is one of the most important

elements of any business continuity plan and disaster recovery plan. System administrators are usually familiar with various types of backups. So, if you bring one or
more individuals with specific technical expertise in this area onto the BCP/DRP
team to provide expert guidance, you can benefit a lot.

There are three main types of backups:

Full Backups: As the name implies, full backups store a complete copy of the data
contained on the protected device. Full backups duplicate every file on the system
regardless of the setting of the archive bit. Once a full backup is complete, the archive
bit on every file is reset, turned off, or set to 0.

Incremental Backups: In incremental backups, only files that have been modified
since the time of the most recent full or incremental backup are stored. Once an
incremental backup is complete, the archive bit on all duplicated files is reset, turned
off, or set to 0.

Differential Backups: Differential backups store all files that have been modified
since the time of the most recent full backup. Only files that have the archive bit
turned on, enabled, or set to 1 are duplicated. However, unlike full and incremental
backups, the differential backup process does not change the archive bit.

The time needed to restore data in the event of an emergency is the most important differentiating factor between incremental and differential backups. If you use
a combination of full and differential backups, you will need to restore only two
backups—the most recent full backup and the most recent differential backup. On
the other hand, you will need to restore the most recent full backups as well as all
incremental backups performed since those full backups if your strategy combines
full backups with incremental backups. The trade-off is the time required to create
the backups. Although differential backups don't take much time to restore, they take
longer to create than incremental ones.

The storage of backup media is equally critical. To easily fulfill user requests for
backup data, it's convenient to store backup media in or near the primary operations
center. However, you'll need to keep copies of the media in at least one offsite
location to provide redundancy should your primary operating location be suddenly
destroyed.

A backup strategy utilizing more than one of the three backup types along with
a media rotation scheme is usually adopted by organizations and both allow backup
administrators access to a sufficiently large range of backups to complete user
requests and provide fault tolerance while minimizing the amount of money that
must be spent on backup media. A common strategy is to perform incremental or
differential backups on a nightly basis and full backups over the weekend. The
specific method of backup and all of the particulars of the backup procedure are
dependent on your organization's fault tolerance requirements. If you are unable to
survive minor amounts of data loss, your ability to tolerate, faults are low. However,
if hours or days of data can be lost without serious consequence, your tolerance of
faults is high. You should design your backup solution accordingly.

Backup Tape Formats

Two factors that a worthwhile backup solution should track and manage are the physical characteristics and the rotation cycle. The type of tape drive in use is a part of the physical characteristics. This defines the physical wear placed on the media. The frequency of backups and retention length of protected data are together known as the rotation cycle. You can be assured that valuable data will be retained on serviceable backup media by observing these characteristics. Backup media has a maximum use limit; perhaps 5, 10, or 20 rewrites may be made before the media begins to lose reliability (statistically speaking). A wide variety of backup tape formats exist:

- Digital Data Storage (DDS)/Digital Audio Tape (DAT)
- Digital Linear Tape (DLT) and Super DLT
- Linear Tape-Open (LTO).

Disk-to-Disk Backup

The cost of disk storage has drastically decreased over the past decade. At the same time, its capability to hold data has increased significantly. Therefore, tape and optical media can no longer cope with data volume requirements anymore.

Many enterprises now use disk-to-disk (D2D) backup solutions for some portion of their disaster recovery strategy. Organizations seeking to adopt an entirely disk-to-disk approach must remember to maintain geographical diversity. Some of those disks have to be located offsite. To solve this problem, many organizations hire managed service providers to manage remote backup locations.

Backup Best Practices

No matter whether you are using the media or the method backup solution, you must address several common issues with backups. For instance, backup and restoration activities can be bulky and slow. So, the performance of a network can be significantly affected when such data movement is being done, especially during regular production hours. Thus, low peak periods should be used to perform this task.

Over time the number of backup data increases. This causes the backup and restoration processes to take longer each time and to consume more space on the backup media. To handle a reasonable amount of growth over a reasonable amount of time into your backup solution, you need to build sufficient capacity.

There is always a chance for data loss every time a periodic backup is run. Murphy's law dictates that a server never crashes immediately after a successful backup. Instead, it is always just before the next backup begins. Some form of real-time continuous backups, such as RAID, clustering, or server mirroring, needs to be deployed to avoid this problem with periods.

Finally, the organization's recovery processes should be tested from time to time. Many organizations often completely rely on the report of their backup software. And if the software reports a successful backup, they take it for granted without running any test. This is one of the biggest causes of backup failures.

Recovery Versus Restoration

Separating disaster restoration tasks from disaster recovery tasks can sometimes be very useful. This is especially true when a recovery process is expected to be very time-consuming. A disaster recovery team may be assigned to implement and maintain operations at the recovery site, and a salvage team is assigned to restore the primary site to operational capacity. These allocations need to be made based on the needs of your organization and the types of disasters you face.

The recovery team members have a very short time frame in which to operate. They must put the DRP into action and restore IT capabilities as swiftly as possible. If the recovery team fails to restore business processes within the MTD/RTO, then the company fails.

The salvage team members begin their work as soon as the original site is deemed safe for people. They work on restoring the company to its full original capabilities as well as to the original location is needed. A new primary spot is selected if the original location doesn't exist anymore. The salvage team must rebuild or repair the IT infrastructure. Since this activity is the same as building a new IT system, the return activity from the alternate or recovery site to the primary or original site is itself a risky activity. Fortunately, the salvage team has more time to work than the recovery team. Once the repair or rebuild process is completed, the salvage team must ensure the reliability of the new IT infrastructure. The stress test can rebuild the network by returning the least mission-critical processes to the restored original state. More important processes are transferred as the restored site shows resiliency. A serious vulnerability exists when mission-critical processes are returned to the original site. The act of returning to the original site could cause a disaster of its own. Therefore, the state of emergency can only be declared once full normal operations have returned to the restored original site.

After any disaster recovery effort, the time will come to restore operations at the primary site and terminate any processing sites operating under the disaster recovery agreement. Your DRP should specify the criteria used to determine when it is appropriate to return to the primary site and guide the DRP recovery and salvage teams through an orderly transition.

Assignment Questions

1.   What are the main techniques used to create offsite copies of database content? Explain?
2.   Explain various backup tape formats
3.   Write a note on Recovery vs. Restoration
4.   Briefly explain Backups and Offsite Storage

Multiple Choice Questions

1.   Activities of an organization such as process and track operations, logistics, and sales are greatly dependent on its ———

     (A) Audit (B) Review (C) Database

2. In electronic vaulting, how the database backups are moved to a remote site

    (A) bulk transfers (B) backups (C) recovery
3. ——— store a complete copy of the data contained on the protected device

    (A) Differential backup (B) incremental backup (C) full backups
4. DDS stands for ....

    (A) Digital Data Service (B) Digital Data Storage (C) Digital Data Setup
5. The type of tape drive in use is a part of the physical characteristics, True or False

    (A) True (B) False (C) None

Answers

1. Database
2. Bulk transfers
3. Full backups
4. Digital data storage
5. True

Summary Questions

1. Explain the three main techniques that are used to create offsite copies of the database content
2. Discuss various types of backups
3. Compare Recovery and Restoration
4. Write a short note on backup tape formats
5. Explain Backups and Offsite Storage

Chapter 17
Analyze Test Output and Generate a Report

Implementing Security Management Processes

Along with performing assessments and testing, many useful information security programs also include various management processes designed to check the effective operation of the information security program. As these processes have a deterrent effect against insider attacks and provide management oversight, they are a critical feedback loop in the security assessment process.

The security management reviews that fill this need include log reviews, account management, backup verification, and essential performance and risk indicators.

Log Reviews

Storing log data and conducting automated and manual log reviews have significant importance in the security management process. Security incident and event management (SIEM) packages play an essential role in these processes, automating much of the routine work of log review. However, to ensure that users are not abusing their privileges, information security managers should also periodically conduct log reviews, particularly for sensitive functions. For example, if an information security team has access to eDiscovery tools that allow searching through the contents of individual user files, security managers should routinely review the logs of actions taken by those administrative users to ensure that their file access relates to legitimate eDiscovery initiatives and does not violate user privacy.

Account Management

Account management reviews ensure that unauthorized modifications do not occur, and users only retain authorized permissions. Account management reviews may be a function of information security management personnel or internal auditors.

Conducting a full review of all user accounts is one of the ways you can perform account management. As this process requires a lot of time to complete, it is usually done for highly privileged accounts.

The exact process may vary from organization to organization, but here's one example:

© The Author(s), under exclusive license to Springer Nature Singapore Pte Ltd. 2023
B. S. Rawal et al., *Cybersecurity and Identity Access Management*,
https://doi.org/10.1007/978-981-19-2658-7_17

- System administrators are asked by managers to provide a list of privileged access and privileged access rights. They may monitor the administrator as they retrieve this list to avoid tampering.
- The privilege approval authority is then asked to provide a list of authorized users and the privileges they should be assigned.
- The managers then compare the two lists to ensure that only authorized users retain access to the system and that the access of each user does not exceed their authorization.

This process may include many other checks, such as verifying that terminated users do not retain access to the system, checking the paper trail for specific accounts, or other tasks. Organizations that do not have time to conduct this thorough process may use sampling instead. In this approach, managers perform a full verification of the method used to grant permissions for those accounts on a randomly picked sample of accounts. If no significant flaws are found in the sample, they assume that this represents the entire population.

Backup Verification

In Chap. 8, "Recovery Strategies for Database," we have discussed the importance of maintaining a consistent backup program. To ensure that the backup process functions appropriately and meets the organization's data protection needs, managers should periodically inspect the results of backups. This inspection may involve reviewing logs, checking hash values, or requesting an actual restore of a system or file.

Key Performance and Risk Indicators

Key performance and risk indicators should also be monitored by security managers on an ongoing basis. The exact metrics they monitor will vary from organization to organization but may include the following:

- Number of open vulnerabilities
- Number of compromised accounts
- Time to resolve vulnerabilities
- Repeat audit findings
- Number of software flaws detected in preproduction scanning
- User attempts to visit known malicious sites.

Once an organization decides which key security metrics they want to track, managers may develop a dashboard that displays the values of these metrics over time and display it where both managers and the security team will regularly see them.

Security Audits and Reviews

Security audits and reviews ensure the implementation of security controls in an organization properly. The effectiveness of access controls is assessed by access review audits. These reviews ensure that accounts don't have excessive privileges, are managed appropriately, and are disabled or deleted when required. On behalf of the Security Operations domain, security audits ensure keeping management controls in place. The following list includes some common items to check:

Patch Management: As soon as the patches are available, a patch management review ensures that these are evaluated properly. It also ensures that the organization follows established procedures to evaluate, test, verify, approve, and deploy the patches. In any patch management review or audit, vulnerability scan reports can be valuable.

Vulnerability Management: Vulnerability scans and assessment performance are reviewed regularly by vulnerability management to ensure that it is in line with compliance with established guidelines. For example, an organization may have a policy document stating that vulnerability scans are performed at least weekly, which is verified by the review that this is done. Additionally, the vulnerabilities discovered in the scans that have been addressed and mitigated will be verified by the review.

Configuration Management: Systems can be audited periodically to ensure that the original configurations remained unmodified. Often scripting tools are used to check specific configurations of systems and identify if a change has occurred or not. Additionally, to record configuration changes, logging can be enabled for many configuration settings. A configuration management audit can check the logs for any change to verify their authorization.

Change Management: A change management review ensures that changes are implemented according to the organization's change management policy. This often includes a cause determination review of outages. Outages that result from unauthorized changes indicate clearly that the change management program needs improvement.

Reporting Audit Results

The actual formats, which vary greatly, are used by an organization to produce reports from audit trails. However, reports should address a few central or basic concepts:

- The scope of the audit
- The purpose of the audit
- The results discovered or revealed by the audit.

In addition to these basic concepts, audit reports often include many details specific to the environment, such as time, date, and a list of the audited systems. They can also include a wide range of content focusing on

- Events, problems, and conditions
- Standards, criteria, and baselines
- Reasons, causes, effect, and impact
- Recommended safeguards and solutions.

Audit reports should have a design or structure that is concise, clear, and objective. Although auditors will often include recommendations or options, they should properly identify them. The actual findings should be based on facts and evidence gathered from audit trials and other sources during the audit.

Protecting Audit Results

Audit reports cover sensitive information. Reports should be assigned a classification label permitting access to only those people with sufficient privilege to audit reports. This includes security personnel and high-level executives involved in the creation of the reports or responsible for the correction of items mentioned in the reports.

Sometimes a separate audit report is created with limited data for other personnel. This modified report includes only the details relevant to the target audience. For example, senior management does not need to know all the tiny details of an audit report. Therefore, the audit report is much more concise and offers more of an overview or summary of findings for senior management. An audit report for a security administrator should be very detailed and includes all available information on the events it covers as he is responsible for correcting the problems.

On the other hand, the fact that an auditor performs an audit is often very publicly. This lets personnel know that senior management is actively taking steps to maintain security.

Distributing Audit Reports

After completion of an audit report, auditors submit it to defined recipients in the security policy documentation. It's common to confirm the receipt by signing. When an audit report contains information about performance issues or serious security violations, personnel escalate it to higher levels of management for notification, review, and assignment of a response to resolve the issues.

Using External Auditors

Many organizations hire external security auditors to conduct independent audits. Additionally, external audits are required by some laws and regulations require. External audits bring a fresh, outside perspective to internal policies, practices, and procedures and provide a level of objectivity that an internal audit cannot provide.

Many organizations hire external security experts as a form of testing to carry out penetration testing against their systems. Penetration tests help an organization to identify vulnerabilities and the ability of attackers to take advantage of these vulnerabilities.

To inspect appropriate aspects of the IT and physical environment, an external auditor is given access to the company's security policy and authorization. Thus, the auditor must be a trusted entity. The audit activity is carried out to obtain a final report that details findings and suggests countermeasures in need.

An external audit requires a considerable amount of time to complete—weeks or, in some cases, months. The auditor may issue interim reports while the audit is still running. An interim report is a verbal or written report given to the organization about any observed policy/procedure mismatches or security weaknesses demanding immediate attention. Auditors issue interim reports whenever a problem or issue is too important to wait until the final audit report.

The auditors typically hold an exit conference after the completion of their investigations. The auditors present and discuss their findings and discuss resolution issues with the affected parties at this conference. However, the auditors write and submit their final audit report to the organization only after the exit conference is over and left the premises after that. This allows the final audit report to remain unaffected by coercion and office politics.

The organizations' internal auditors review the final audit report and make recommendations to senior management based on the report. Senior management carries out the Herculean task of selecting which recommendations to implement and for offering implementation requirements to internal personnel.

Assignment Questions

1. Explain the role of log review in security management
2. Write a note on account management
3. Define Vulnerability Management
4. List out and explain the basic concepts a report should address
5. Explain the importance of external audits

Multiple Choice Questions

1. SIEM stands for?

   (A) Secure Idea Establishment and Management (B) Security incident and event management (C) Secure Incident and Event Marketing
2. Account management is usually done for

   (A) Normal accounts (B) privileged accounts (C) highly privileged accounts
3. Backup verification involves the following steps

   (A) Reviewing logs (B) checking hash values (C) Both a and b
4. The effectiveness of access controls is assessed by access

   (A) Review audits (B) Security audits (C) None of the above
5. The auditors typically hold ——— after completion of their investigations

   (A) A complete report (B) An exit conference (C) A review meeting

Answers

1. Security incident and event management
2. Highly privileged accounts
3. Both a and b
4. Review audits
5. An exit conference

Summary Questions

1. What do you mean by account management? Explain the process with an example
2. Explain the importance of log reviews
3. Explain the metrics the security managers use to monitor the key performance and risk indicators

Chapter 18
Ensure Appropriate Asset Retention

An asset security domain, throughout its life cycle, focuses on handling, collecting, and protecting information. This domain, as a primary step, classifies information based on its value to the organization. Depending on the classification, all follow-on actions vary. For example, unclassified data uses fewer security controls, whereas highly classified data requires stringent security controls.

Classifying and Labeling Assets

One of the first steps in asset security is labeling and classifying the assets. Often within a security policy, classification definitions are included by the organization, and then based on the security policy requirements, personnel label assets appropriately. In this context, assets include sensitive data, the media used to hold it, and the hardware used to process it.

Data to Retain

The data relating to partnership, third-party dealings, or business management is valuable for any organization. Moreover, the counsel's opinion has paramount importance over administrators because he suggests what data is useful in the event of litigation.

Defining Sensitive Data

Sensitive data is any information that is classified or isn't public. It includes proprietary, confidential, protected, or any other type of data that an organization deems to protect considering its value to the organization or to comply with existing regulations and laws.

Personal Data

Personal data is any information relating to an identifiable person. Institute of Standards and Technology (NIST) Special Publication (SP) 800-122 provides a more formal definition.

© The Author(s), under exclusive license to Springer Nature Singapore Pte Ltd. 2023 215
B. S. Rawal et al., *Cybersecurity and Identity Access Management*,
https://doi.org/10.1007/978-981-19-2658-7_18

"Any information about an individual maintained by an agency, including

(1)   any information that can be used to distinguish or trace an individual's identity, such as name, social security number, date, and place of birth, mother's maiden name, or biometric records; and

(2)   any other information that is linked or linkable to an individual, such as medical, educational, financial, and employment information."

The key is that the proper protection of PII is the organizations' liability. This includes PII related to customers and employees. When a data breach results in a compromise of PII, many laws require organizations to notify individuals.

Protected Health Information

Protected health information (PHI) is any information that can be related to specific persons' health issues. In the USA, the Health Insurance Portability and Accountability Act (HIPAA) mandates the protection of PHI. HIPAA provides a more formal definition of PHI:

Health information means any sort of health care information, whether recorded or oral in any form or medium, that—

(A)   is received or created by a healthcare provider, public health authority, life insurer, health plan, employer, school, or university, or healthcare clearinghouse; and

(B)   relates to the present, past, or future mental or physical health or condition of any individual, the provision of health care to an individual, or the present, past, or future payment for the provision of health care to an individual."

Some people think that it's only the medical care providers such as doctors and hospitals who are bound to protect PHI. However, the definition of PHI by HIPAA is much broader. Any employer that supplements or provides healthcare policies handles and collects PHI. Most commonly, organizations provide or supplement healthcare policies, so HIPAA applies to a large percentage of organizations in the USA.

Proprietary Data

Proprietary data contains any data that helps an organization maintaining a competitive edge. It could be software code it developed, internal processes, technical plans for products, intellectual property, or trade secrets. It can seriously affect the primary mission of an organization if competitors can access proprietary data.

Although patents, copyrights, and trade secret laws provide a level of protection for proprietary data, this isn't always enough. Many criminals don't pay attention to patents, copyrights, and laws. Similarly, a significant amount of proprietary data has been stolen by foreign entities.

As an example, Mandiant, an information security company, released a report in 2013 documenting a group operating out of China that they named APT1. Mandiant attributes a significant number of data thefts to this advanced persistent threat (APT). They observed, 141 companies spanning 20 major industries have been compromised by APT1. In one instance, they observed 6.5 TB of compressed intellectual property data was stolen by APT1 over a ten-month period.

Retaining Assets

An asset is anything, like partners, employees, facilities, equipment, and information, which are important to the organization. Information is valuable to every information system and is considered the most important asset to any company or organization. A company's information system transit information within authorized accounts and dispose of it appropriately after it loses its necessity.

Retention requirements are applied to records or data, media holding sensitive data, personnel who have access to sensitive data, and systems that process sensitive data. Media retention and record retention is the most important element of asset retention.

Record retention involves maintaining and retaining important information as long as it is needed, and destroying it loses necessity. Retention timeframes are typically identified by an organization's security policy or data policy. Some regulations and laws bound the length of time that an organization should retain data, such as two years, seven years, or even indefinitely. However, an organization should still identify how long to retain data, though there are no external requirements. To reconstruct past incidents, audit trail data needs to be kept long enough. But the organization must specify the length and depth of investigation. A current trend with many organizations is to cut short retention policies with an email to reduce legal liabilities.

As an example, many organizations require three years or longer for the retention of all audit logs. This allows the organization to reconstruct the details of past security incidents. There is a fear of administrators deleting valuable data earlier than management expects them to submit or attempt to keep data indefinitely when an organization doesn't have a retention policy. The length of data retention determines the costs in terms of media, personnel to protect it, and locations to store it.

Most hardware is on a refresh cycle, where it is replaced every three to five years. Hardware retention primarily refers to the retention of information until it has been properly sanitized. Personnel retention in this context implies the knowledge that personnel gains while working for an organization. It's common for organizations to include nondisclosure agreements (NDAs) when hiring new personnel. The agreements in NDAs prevent employees from sharing proprietary data with others and leaving the job without formal notification.

Protecting Confidentiality with Cryptography

Encryption is one of the primary methods of protecting the confidentiality of data. As an introduction, encryption converts clear text data into scrambled ciphertext. When the data is in cleartext format, anyone can read it. However, it is almost impossible to read the scrambled ciphertext when strong encryption algorithms are used.

Protecting Data with Symmetric Encryption

To encrypt and decrypt data, symmetric encryption uses the same key. In other words, if an algorithm encrypted data with a key of 123, it would decrypt it with the same key of 123. Symmetric algorithms use different keys for different data. For example, if it encrypted one set of data using a key of 123, the next set of data might be encrypted with a key of 456. The important point here is that a file encrypted using a key of 123 can only be decrypted using the same key of 123. Practically, the key size is much larger. For example, AES uses key sizes of 128 bits or 192 bits, and AES 256 uses a key size of 256 bits.

The following list identifies some of the commonly used symmetric encryption algorithms. Although many of these algorithms are used in applications to encrypt data at rest, some of them are also used in transport encryption algorithms.

Advanced Encryption Standard: One of the most popular symmetric encryption algorithms, the Advanced Encryption Standard (AES). Back in 2001, it was a standard replacement for the older Data Encryption Standard (DES) by NIST. Since then, AES is implemented steadily by developers into many other algorithms and protocols. As an example, The Microsoft Encrypting File System (EFS) uses AES for file and folder encryption. BitLocker (a full disk encryption application used with a Trusted Platform Module) uses AES. AES supports key sizes of 128 bits, 192 bits, and 256 bits, and the US government has approved its use to protect classified data up to top secret. Additional security is added by larger key sizes, making it more difficult for unauthorized personnel to decrypt the data.

Triple DES: As a possible replacement for DES, developers created Triple DES (or 3DES). The newer implementations use 112-bit or 168-bit keys, but the first implementation used 56-bit keys. A higher level of security is provided by larger keys provide. System Center Configuration Manager and Microsoft OneNote use 3DES to protect some content and passwords.

Blowfish: As a possible alternative to DES, Security expert Bruce Schneier developed Blowfish. It is a strong encryption protocol and can use key sizes of 32 bits to 448 bits. Bcrypt is based on Blowfish, which is used in Linux systems to encrypt passwords. As a salt, Bcrypt adds 128 additional bits to protect against rainbow table attacks.

Protecting Data with Transport Encryption

Before being transmitted, transport encryption methods encrypt data, protecting data in transit. A sniffing attack is a primary risk of sending unencrypted data over a network. To capture traffic sent over a network, attackers can use a sniffer or protocol analyzer. Attackers can read all the data sent in cleartext allowed by the sniffer. However, attackers are unable to read data encrypted with a strong encryption protocol.

As an example, to encrypt e-commerce transactions, web browsers mostly use Hypertext Transfer Protocol Secure (HTTPS). This prevents attackers from using credit card information and capturing the data to rack up charges. In contrast, the Hypertext Transfer Protocol (HTTP) transmits data in cleartext.

Transport Layer Security (TLS) is used in almost all HTTPS transmissions as the underlying encryption protocol. TLS comes following the Secure Sockets Layer (SSL). Netscape created and released SSL in 1995. Later, TLS was released as a replacement for SSL by the Internet Engineering Task Force (IETF). In 2014, Google discovered that SSL is susceptible to the POODLE attack (Padding Oracle on Down-graded Legacy Encryption). As a result, SSL was disabled by many organizations in their applications.

Remote access solutions like virtual private networks (VPNs) are often enabled in Organizations. VPNs are used to allow employees to access the organization's internal network when they are working from their home or while traveling. Encryption is important as VPN traffic goes over a public network, such as the internet. VPNs use encryption protocols such as TLS and Internet Protocol security (IPsec).

Often Layer 2 Tunneling Protocol (L2TP) for VPNs combine with IPsec. Though L2TP transmits data in cleartext, during the transition, L2TP/IPsec encrypts data and sends it over the internet using tunnel mode to protect it. IPsec includes Encapsulating Security Payload (ESP) to provide confidentiality and an Authentication Header (AH) to provide authentication and integrity.

Before transmitting sensitive data on internal networks, it's appropriate to encrypt, and Secure Shell (SSH) and IPsec are commonly used to protect data in transit on internal networks. SSH is a strong encryption protocol, including other protocols like Secure File Transfer Protocol (SFTP) and Secure Copy (SCP). Both SFTP and SCP are secure protocols used to transfer encrypted files over a network. Protocols such as File Transfer Protocol (FTP) are not appropriate for transmitting sensitive data over a network as it transmits data in cleartext.

When administering remote servers, many administrators use SSH instead of Telnet. Telnet sends traffic over the network in cleartext, so it should not be used. Also, Telnet sends their credentials over the network in cleartext, which is why administrators need to log on to the server when connecting to remote servers. However, SSH encrypts all of the traffic, including the administrator's credentials.

Retention Policies

Data protection requires that sensitive data should not be preserved for a longer time when processed for any purpose. Unfortunately, there is no universal agreement on how long the organization should retain data. However, the legal and regulatory requirements vary among regions, countries, and business communities. Data retention policies must be followed by every organization to thwart disaster, especially when coping with ongoing or pending litigations.

Examples of retention policies include:

- The State of Florida Electronic Records and Records Management Practices, 2010
- The European Documents Retention Guide, 2012.

Developing a Retention Policy

Every retention policy must answer three fundamental questions, which are:

How to Retain Data: The data should be kept in an accessible manner so that, whenever required, it can be accessed. The organization should consider some issues to make this accessibility certain, including:

- The Taxonomy is a data classification scheme. This classification involves various categories, including the organizational (executive, union employee), the functional (human resource, product developments), or any combination of these.
- The Normalization develops tagging schemes to ensure that the data is searchable. Non-normalized data is kept in various formats such as PDF files, audio, and video.

Data Retention Period

The classical data retention longevity approaches were: "the keep nothing" camp and "the keep everything" camp. But these approaches are dysfunctional in modern times under many circumstances, particularly when an organization encounters a lawsuit. As stated before, there is no universal pact on data retention policies.

Assignment Questions

1. According to NIST, how a personal data is defined?
2. Every retention policy must answer three fundamental questions. What are they?
3. The Taxonomy is a data classification scheme; justify your answer
4. Give some examples for retention policies
5. Explain Protecting Data with Symmetric Encryption

Multiple Choice Questions

1. What are the first steps in asset security?

    (A) labeling and classifying the assets (B) defining sensitive data (C) defining personal data
2. There is no universal pact on data retention policies. True or false

    (A) True (B) False
3. ——— is the primary risk of sending unencrypted data over a network.

    (A) encoding (B) encryption (C) Sniffing attack
4. Transport Layer Security (TLS) is used in almost all HTTPS transmissions as the underlying encryption protocol. True or False

    (A) True (B) False
5. AES 256 uses a key size of ——— bits.

    (A) 192 (B) 128 (C) 256

Answers

1. Labeling and classifying the assets
2. True
3. Sniffing attack
4. True
5. 256

Summary Questions

1. Define Personal data as defined by NIST
2. Explain Protecting Data with Symmetric Encryption
3. Every retention policy must answer three fundamental questions. What are they?
4. Explain retention policies with examples
5. Explain Protecting Data with Transport Encryption

Chapter 19
Determine Information and Security Controls

Defining Information Classification

Data owners define data classifications and ensure systems and are responsible for marking data properly. A data classification recognizes the value of the data to the organization and is critical to protect data integrity and confidentiality. Additionally, data owners explain requirements to protect data at different classifications, such as encrypting sensitive data in transit and at rest. Data classifications are typically defined within security policies or data policies. The essential metadata items are a classification level that is attached to organizations' valuable information. The classification tag ensures the protection of information and remains affixed throughout the information life cycle of system (Acquisition, Use, Archival, and Disposal).

The words used to classify information are "sensitivity," "criticality," sometimes in combination. If unauthorized individuals access information, the "sensitivity" of it is compromised. For example, the information losses suffered by the organizations, such as the Office of Personnel Management and the National Security Agency. On the other hand, to function properly, any organization requires critical information. For example, Code Spaces, a company offering code repository services, was shut down in 2014 when unauthorized individuals deleted their code repositories.

The organization can choose the classification level depending on whether it has a military agency or commercial business. The classification rule must be applied to data irrespective of its format; it doesn't matter whether the data is digital, audio, video, fax, paper, etc. The typical levels of commercial business and military data involve:

- Public data can be viewed by the general public, and, therefore, the exposure of this data could not cause any damage. For example, the general public can be aware of the government's upcoming financial aid projects.
- Sensitive information is any type of classified information, and proper management requires to prevent unauthorized manifestation resulting in a loss of confidentiality. Sensitive information needs extraordinary precautions to ensure integrity and confidentiality for its protection. Proper management includes marking,

B. S. Rawal et al., *Cybersecurity and Identity Access Management*,
https://doi.org/10.1007/978-981-19-2658-7_19

handling, storing, and destroying sensitive information. The two areas where organizations often miss the mark are sanitizing media or equipment when it is at the end of its life cycle and adequately protecting backup media storing sensitive information. For example, sensitive data may comprise the company's financial information.

- Private data may comprise personal information, such as bank accounts and credit card information. Unauthorized disclosure can be disastrous.
- Confidential information: The confidential label is "applied to information, the unauthorized disclosure of which reasonably could be expected to cause damage to the national security that the original classification authority is able to identify or describe."
- Unclassified information: Unclassified refers to any data that doesn't meet one of the descriptions for secret or confidential, secret, or top secret data, such as information on recruitment in the military.
- Secret information: The secret label is "applied to information, the unauthorized disclosure of which reasonably could be expected to cause serious damage to the national security that the original classification authority is able to identify or describe," such as the release of military deployment plans.
- Top secret information: The top secret label is "applied to information, the unauthorized disclosure of which reasonably could be expected to cause exceptionally grave damage to the national security that the original classification authority is able to identify or describe," such as the disclosure of spy satellite information.

Non-government organizations hardly need to classify their potential database threat to national security. However, management is worried about the potential threat to the organization. Some non-government organizations use labels such as Class 0, Class 1, Class 2, and Class 3. Other organizations utilize more meaningful labels such as public, private, sensitive, and confidential (or proprietary). Figure 19.1 shows the relationship between these different classifications with the non-government (or civilian) classifications on the right and the government classifications on the left. Just as the government can define the data based on the potential adverse impact of a data breach, organizations can use similar descriptions.

Identifying Data Roles

The completion of the life cycle successfully is a must for the transition of information. The diversified entities that make the life cycle successful include the data custodian, data owners, system owner, supervisor, security administrator, and user. Each role has uniqueness in protecting the organization's assets.

The Data Owner, or Information Owner, is a manager who determines the classification level and ensures data protection. He also determines whether the data is in soft-copy or hard-copy form.

The data owner is the person having ultimate organizational responsibility for data. The owner is typically a department head (DH), the CEO, or the president. Data owners identify the classification of data and ensure that it is labeled accordingly. They also ensure the organization's security policy per requirements and adequate

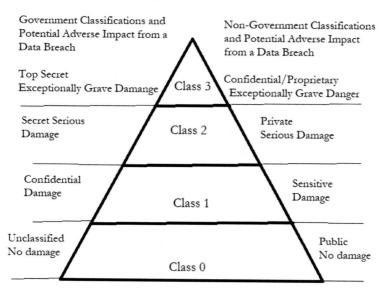

Fig. 19.1 Data classifications

security controls based on the classification. If they fail to perform due diligence in establishing and enforcing security policies, owners may be liable for negligence to protect and sustain sensitive data.

- Same as the data owner, NIST SP 800-18 configures the following responsibilities for the information owner, which can be interpreted as.
- Establishes the rules for appropriate protection and use of the subject data/information (rules of behavior).
- Provides security controls for the information system(s) where the information resides and input to information system owners regarding the security requirements.
- Decides who has access to the information system and with what types of privileges or access rights.
- Assists in the identification and assessment of the common security controls where the information resides.

The system owner controls the working of the computer that stores data. This involves the software and hardware configurations, such as managing system updates, patches, and so on.

The data custodian performs frequent data backups and restoration and maintains security, such as the configuration of antivirus programs. Data owners often delegate day-to-day tasks to a custodian. A custodian helps protect the integrity and security of data by ensuring it is properly stored and protected. For example, custodians would ensure the data is backed up in accordance with a backup policy. If administrators have configured auditing on the data, custodians would also maintain these logs.

In practice, personnel within an IT department or system security administrators would typically be the custodians. They might be the same administrators responsible for assigning permissions to data.

The security administrator, a data administrator, is responsible for granting appropriate access to personnel as he assigns permission and handles data on a network. Though they don't necessarily have full administrator rights and privileges, they do have the ability to assign permissions. Administrators granting users access to only what they need for their job and assign these permissions based on the principles of least privilege they need to have.

Administrators typically assign permissions using a role-based access control model. In other words, they add user accounts to groups and then grant permissions to the groups. When users no longer need access to the data, administrators remove their accounts from the group.

The user, a user, is any person who accomplishes work tasks by accessing data via a computing system. Users can only access the data they need to perform their work tasks. You can also think of users as end-users or employees. Users must comply with mandatory policies, rules, standards, and procedures. For instance, the user should not share any confidential information or his account with other colleagues.

The Supervisor, or User Manager, is responsible for overseeing the activities of all the entities stated above.

Protecting Privacy

Many countries moved toward security instead of privacy after the traumatic attack of 9/11 in New York City. However, the countries were motivated more to focus on more privacy protection after the security leaks of Edward Snowden in 2013. Many organizations consider both privacy and security in their information systems.

Organizations have a liability to protect the data that they collect and maintain. This is especially true for both PII and PHI data. Many laws and regulations mandate the protection of privacy data, and organizations have a responsibility to learn about laws and regulations to apply them properly. Additionally, organizations need to check thoroughly to ensure their practices comply with these laws and regulations.

Many laws require organizations to disclose what data they possess, what they collect, why they collect it, and how they plan to use the information. In addition, these laws prohibit organizations from using the information in ways that are outside the scope of what they intend to use it for. As an example, the organization should not sell the email addresses to third parties if an organization states it is collecting email addresses to communicate with a customer about purchases.

Commonly organizations use an online privacy policy on their websites to protect them from breaches. Some of the entities that require strict adherence to privacy laws include

- The USA (with HIPAA privacy rules)
- The state of California (with the California Online Privacy Protection Act of 2003)
- Canada (with the Personal Information Protection and Electronic Documents Act)
- EU with the General Data Protection Regulation (GDPR).

The scheduled timeline is for organizations to begin adopting the requirements in 2015 and 2016 and begin enforcing the requirements in 2017 and 2018.

If organizations operate in the jurisdiction of the law, many of these laws require them to follow these requirements. For example, the California Online Privacy Protection Act (COPA) requires a conspicuously posted privacy policy for any online services or commercial websites to collect personal information on California residents. Practically, this potentially applies to any website in the world that collects personal information because if the website is accessible on the internet, any California residents can access it. COPA is considered to be one of the most stringent laws in the USA, and US-based organizations that follow the requirements of the California law typically meet the requirements in other locales. However, an organization has an obligation to determine what laws apply to it and follow them.

An organization will typically use several different security controls to protect privacy. Proper selection of security controls can be a daunting task, especially for new organizations. However, identifying standards and using relevant security baselines make the task a little easier.

Data owners play an important role in privacy protection as they indirectly or directly decide who has access to particular data.

Data remnants are still left even after the owner deletes the data, and they could badly threaten privacy. In fact, without erasing the original data, the data deletion operation just marks the memory available for other data. There are four approaches used to counter data remanence:

- Overwriting replaces original data memory location making it unrecoverable (the pattern of 0's and 1's) with the random or fixed patterns of 0's and 1's.
- Degaussing uses a magnetic force to remove the magnetic field patterns on disk drives. As a result, the original data is wiped to be unrecoverable.
- In encryption, the key to the data is only available to the owner of the data, which makes the data unusable even after deletion.
- The physical media is destroyed using a shredding technique to achieve physical destruction.

Limits on Collection: A minimum amount of data must be collected by an organization because it can be a matter of law later on. In 2014, more than 100 countries passed privacy protection laws that affect organizations in their jurisdictions. The privacy protection policies vary among countries; for example, China has no restrictions, while Argentina has the most restrictive privacy.

Data Security Controls

Determining data security controls is a pretty difficult task. However, in scoping and tailoring, the standards are employed to choose the controls. Also, the control's determination is affected by the situation, either the data is in use, in motion, or at rest. Figure 19.2 shows the states of data.

Scoping and Tailoring: Scoping determines which standard will be used by the organization. Scoping refers to the review of baseline security controls and the selection of only those controls that apply to the IT system you're trying to protect. As an

Fig. 19.2 States of data

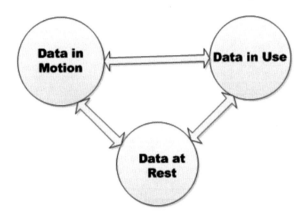

Table 19.1 Insecure network protocols and reliable solutions

Action	Not appropriate	Appropriate
Web access	HTTP	HTTPS
File transfer	FTP, RCP	FTPS, *SFTP, SCP*
Remote desktop	VNC	RDP, radmin
Remote shell	telnet	SSH3

example, there's no need to apply a concurrent session control if a system doesn't allow any two people to log on to it at the same time.

Tailoring is a process referring to modify the list of security controls within a baseline so that they align with the mission of the organization. The tailoring helps to customize the standard for organizations. For example, though an organization might decide that a set of baseline controls applies perfectly to computers in their main location, some controls aren't appropriate or feasible in a remote office location. The organization can select compensating security controls to tailor the baseline to the remote location to handle this situation.

Data that is being transmitted across the network is called data in motion, while data is being stored on the hard drive is data at rest. Each of them needs unique controls for protection.

Drive Encryption is the regulation for the protection of data at rest. This control is recommended for all cellular devices and media that contain confidential information.

Media Transportation and Storage provide data protection through off-site data storage facilites and backup through physical movement or via networks.

Protecting data in motion requires the secured transition of data via networks.

Table 19.1 contains examples of insecure network protocols and reliable solutions:

Using Security Baselines

Baselines provide a starting point, ensuring a minimum-security standard. Imaging is one common baseline that organizations use. Configuring a single system with

Table 19.2 Security control baselines

Control name	Control no	Priority
Access control policy and procedures	AC-1	P-1
Account management	AC-2	P-1
Security awareness training	AT-2	P-1
Separation of duties	AC-5	P-1
Least privilege	AC-6	P-1
Unsuccessful login attempts	AC-7	P-2
Concurrent session control	AC-10	P-3

desired settings, administrators capture it as an image and then deploy the image to other systems ensuring all of the systems are deployed in a similar secure state.

After administrators deploy systems in a secure state, they audit the processes periodically to check the systems to ensure they remain in a secure state. For example, Microsoft Group Policy can periodically check systems and reapply settings to match the baseline.

NIST SP 800-53 discusses security control baselines as a list of security controls. It stresses that a single set of security controls is not applicable for all situations, but any organization can select a set of baseline security controls and tailor it to its needs. Four prioritized sets of security controls are included in Appendix D of SP 800-53 that organizations can implement to provide basic security giving organizations insight into what they should implement first, second, and last.

As an example, consider Table 11.2, which is a partial list of some security controls in the access control family. NIST has assigned the control name and the control number for these controls and has provided a recommended priority. P-1 indicates the highest priority, P-2 is next, and P-3 is last. In Appendix F, NIST SP 800–53 explains all of these controls in more depth (Table 19.2).

It's worth noting that many of the items are labeled for basic security practices as P-1. Access control procedures and policies ensure that users have unique identifications (such as usernames) and can prove their identity with authentication procedures. Based on their proven identity (such as authorization processes), administrators grant users access to resources. Similarly, the implementation of basic security principles such as separation of duties and the principle of least privilege shouldn't be a surprise to anyone studying for the CISSP exam. Of course, it doesn't mean organizations implement them just because these are basic security practices. Unfortunately, many organizations have yet to enforce or discover the basics.

Assignment Questions

1. Explain the typical levels of commercial business and military data
2. Explain the following
     (a) sensitive information (b) confidential information (c) unclassified information
3. How are data classified? Explain

4. Give some examples for some entities that require strict adherence to privacy laws
5. Explain the four approaches used to counter data remanence.

Multiple Choice Questions

1. Data classifications are typically defined within security policies or data policies. True or False

    (A) True (B) False
2. The words used to classify information are "sensitivity," and "———,"

    (A) Confidential (B) encryption (C) criticality
3. GDPR stands for

    (A) General Data Protection Regulation (B) Global Data Protection Regulation (C) General Disk Protection Regulation
4. How many states are there for data?

    (A) 2 (B) 3 (C) 4
5. Drive Encryption is the regulation for the protection of data at rest. True or False

    (A) True (B) False

Answers

1. True
2. Criticality
3. General Data Protection Regulation
4. 3
5. True

Summary Questions

1. Define the following

    (a) sensitive information (b) confidential information (c) unclassified information
2. Explain the typical levels of commercial business and military data
3. How are data classified? Explain
4. List out some entities that require strict adherence to privacy laws
5. Explain the four approaches used to counter data remanence

References

1. Wong A, Yeung A (2009) Introduction to network infrastructure security. In: Network infrastructure security. Springer, Boston, MA. https://doi.org/10.1007/978-1-4419-0166-8_1
2. Kabir MF, Hartmann S (2018) Cybersecurity challenges: an efficient intrusion detection system design. In: 2018 international young engineers forum (YEF-ECE), Costa da Caparica, pp 19–24. https://doi.org/10.1109/YEF-ECE.2018.8368933
3. Alshammari A, Alhaidari S, Alharbi A, Zohdy M (2017) Security threats and challenges in cloud computing. In: 2017 IEEE 4th international conference on cyber security and cloud computing (CSCloud), New York, NY, pp 46–51. https://doi.org/10.1109/CSCloud.2017.59
4. Scarfone K, Mell P (2007) Guide to intrusion detection and prevention systems (IDPS). NIST Spec Publ 800:94
5. Buczak AL, Guven E (2016) A survey of data mining and machine learning methods for cybersecurity intrusion detection. IEEE Commun Surv Tutor IEEE 18:1153–1176
6. Wong K, Dillabaugh C, Seddigh N, Nandy B (2017) Enhancing Suricata intrusion detection system for cybersecurity in SCADA networks. In: 2017 IEEE 30th Canadian conference on electrical and computer engineering (CCECE), pp 1–5
7. Schmittner C, Macher G (2019) Automotive cybersecurity standards—relation and overview. In: Romanovsky A, Troubitsyna E, Gashi I, Schoitsch E, Bitsch F (eds) Computer safety, reliability, and security. SAFECOMP 2019. Lecture notes in computer science, vol 11699. Springer, Cham. https://doi.org/10.1007/978-3-030-26250-1_12
8. Vehicle Electrical System Security Committee (2016) SAE J3061 cybersecurity guidebook for cyber-physical automotive systems. Technical report, SAE
9. Secretary of TF-CS/OTA UNECE WP29 (2019) Draft recommendation on cyber security of the task force on cyber security and over-the-air issues of UNECE WP.29 IWG ITS/AD, Apr 2018. https://wiki.unece.org/pages/viewpage.action?pageId=58524794. Accessed 27 Mar 2019
10. Chockalingam S, Pieters W, Teixeira A, van Gelder P (2017) Bayesian network models in cyber security: a systematic review. In: Lipmaa H, Mitrokotsa A, Matulevičius R (eds) Secure IT systems. NordSec 2017. Lecture notes in computer science, vol 10674. Springer, Cham. https://doi.org/10.1007/978-3-319-70290-2_7
11. Sommestad T, Ekstedt M, Johnson P (2009) Cybersecurity risks assessment with Bayesian defense graphs and architectural models. In: 2009 42nd Hawaii international conference on system sciences, HICSS 2009. IEEE, pp 1–10
12. Ekstedt M, Sommestad T (2009) Enterprise architecture models for cybersecurity analysis. In: Power systems conference and exposition. IEEE, pp 1–6
13. Krishna CGL, Murphy RR (2017) A review on cybersecurity vulnerabilities for unmanned aerial vehicles. In: 2017 IEEE international symposium on safety, security and rescue robotics (SSRR), Shanghai, pp 194–199. https://doi.org/10.1109/SSRR.2017.8088163

© The Editor(s) (if applicable) and The Author(s), under exclusive license to Springer Nature Singapore Pte Ltd. 2023
B. S. Rawal et al., *Cybersecurity and Identity Access Management*,
https://doi.org/10.1007/978-981-19-2658-7

14. Rao A, Carreón N, Lysecky R, Rozenblit J (2018) Probabilistic threat detection for risk management in cyber-physical medical systems. IEEE Softw 35(1):38–43. https://doi.org/10.1109/MS.2017.4541031

15. Labuschagne WA, Burke I, Veerasamy N, Eloff MM (2011) Design of cybersecurity awareness game utilizing a social media framework. In: 2011 information security for South Africa, Johannesburg, pp 1–9. https://doi.org/10.1109/ISSA.2011.6027538

16. Patil S, Jangra A, Bhale M, Raina A, Kulkarni P (2017) Ethical hacking: the need for cybersecurity. In: 2017 IEEE international conference on power, control, signals and instrumentation engineering (ICPCSI), Chennai, pp 1602–1606. https://doi.org/10.1109/ICPCSI.2017.8391982

17. Jaquet-Chiffelle DO, Loi M (2020) Ethical and unethical hacking. In: Christen M, Gordijn B, Loi M (eds) The ethics of cybersecurity. The international library of ethics, law, and technology, vol 21. Springer, Cham. https://doi.org/10.1007/978-3-030-29053-5_9

18. Bahashwan AA, Anbar M, Hanshi SM (2020) Overview of IPv6 based DDoS and DoS attacks detection mechanisms. In: Anbar M, Abdullah N, Manickam S (eds) Advances in cyber security. ACeS 2019. Communications in computer and information science, vol 1132. Springer, Singapore. https://doi.org/10.1007/978-981-15-2693-0_11

19. Javaid A, Niyaz Q, Sun W, Alam M (2016) A deep learning approach for network intrusion detection system. In: Proceedings of the 9th EAI international conference on bio-inspired information and communications technologies (formerly BIONETICS). ICST (Institute for Computer Sciences, Social-Informatics and Telecommunications Engineering), pp 21–26

20. Mousavi SM, St-Hilaire M (2015) Early detection of DDoS attacks against SDN controllers. In: 2015 international conference on computing, networking and communications (ICNC). IEEE, pp 77–81

21. Elejla OE, Anbar M, Belaton B (2017) ICMPv6-based DoS and DDoS attacks and defense mechanisms. IETE Tech Rev 34(4):390–407

22. Alsadhan AA, Hussain A, Baker T, Alfandi O (2018) Detecting distributed denial of service attacks in neighbor discovery protocol using machine learning algorithm based on streams representation. In: Huang DS, Gromiha M, Han K, Hussain A (eds) Intelligent computing methodologies, vol 10956. Springer, Cham, pp 551–563

23. Arjuman NC, Manickam S (2015) A review on ICMPv6 vulnerabilities and its mitigation techniques: classification and art. In: 2015 international conference on computer, communications, and control technology (I4CT). IEEE, pp 323–327

24. Wangen G, Shalaginov A, Hallstensen C (2016) Cyber security risk assessment of a DDoS attack. In: Bishop M, Nascimento A (eds) Information security. ISC 2016. Lecture notes in computer science, vol 9866. Springer, Cham. https://doi.org/10.1007/978-3-319-45871-7_12

25. Wangen G, Hallstensen C, Snekkenes E (2016) A framework for estimating information security risk assessment method completeness—core unified risk framework. Submitted for review

26. Agarwal S, Tyagi A, Usha G (2020) A deep neural network strategy to distinguish and avoid cyber-attacks. In: Dash S, Lakshmi C, Das S, Panigrahi B (eds) Artificial intelligence and evolutionary computations in engineering systems. Advances in intelligent systems and computing, vol 1056. Springer, Singapore. https://doi.org/10.1007/978-981-15-0199-9_58

27. Yuan X, Li C, Li X (2017) DeepDefense: identifying DDoS attack via deep learning. In: IEEE international conference on smart computing (SMARTCOMP)

28. Noh S, Lee C, Choi K, Jung G (2003) Detecting distributed denial of service (DDoS) attacks through inductive training. In: International conference on intelligent data engineering and automated learning. Springer

29. Vyawahare M, Chatterjee M (2020) Taxonomy of cyberbullying detection and prediction techniques in online social networks. In: Jain L, Tsihrintzis G, Balas V, Sharma D (eds) Data communication and networks. Advances in intelligent systems and computing, vol 1049. Springer, Singapore. https://doi.org/10.1007/978-981-15-0132-6_3

30. Kizza JM (2017) Cyber crimes and hackers. In: Guide to computer network security. Computer communications and networks. Springer, Cham. https://doi.org/10.1007/978-3-319-55606-2_5

31. The complete history of hacking. http://www.wbglinks.net/pages/history/
32. Denning D. Activism, hacktivism, and cyberterrorisim: the internet as a tool or influencing foreign policy. http://www.nautilus.og/info-policy/workshop/papers/denning.html
33. Mishra S, Dhir S, Hooda M (2016) A study on cyber security, its issues and cyber crime rates in India. In: Saini H, Sayal R, Rawat S (eds) Innovations in computer science and engineering. Advances in intelligent systems and computing, vol 413. Springer, Singapore. https://doi.org/10.1007/978-981-10-0419-3_30
34. Shah S, Mehtre BM (2015) An overview of vulnerability assessment and penetration testing techniques. J ComputVirol Hack Tech 11:27–49. https://doi.org/10.1007/s11416-014-0231-x
35. Shah S (2013) Vulnerability assessment and penetration testing (VAPT) techniques for cyber defense. IET-NCACNS' SGGS, Nanded
36. Liu B, Shi L, Cai Z (2012) Software vulnerability discovery techniques: a survey. In: IEEE 4th international conference on multimedia information networking and security
37. Al-Ahmad AS, Kahtan H (2019) Fuzz test case generation for penetration testing in mobile cloud computing applications. In: Vasant P, Zelinka I, Weber GW (eds) Intelligent computing and optimization. ICO 2018. Advances in intelligent systems and computing, vol 866. Springer, Cham. https://doi.org/10.1007/978-3-030-00979-3_27
38. Kals S, Kirda E, Kruegel C, Jovanovic N (2006) Secubat: a web vulnerability scanner. In: Proceedings of the 15th international conference on World Wide Web. ACM, pp 247–256
39. Ahmed H, Alsadoon A, Prasad PWC, Costadopoulos N, Hoe LS, Elchoemi A (2017) Next generation cybersecurity solution for an eHealth organization. In: 2017 5th international conference on information and communication technology (ICoIC7), Malacca City, pp 1–5. https://doi.org/10.1109/ICoICT.2017.8074723
40. Salim MM, Rathore S, Park JH (2020) Distributed denial of service attacks and its defenses in IoT: a survey. J Supercomput 76:5320–5363. https://doi.org/10.1007/s11227-019-02945-z
41. Yin D, Zhang L, Yang K (2018) A DDoS attack detection and mitigation with software-defined internet of things framework. IEEE Access 6:24694–24705
42. Vasek M, Thornton M, Moore T (2014) Empirical analysis of Denial-of-Service attacks in the bitcoin ecosystem. In: Böhme R, Brenner M, Moore T, Smith M (eds) Financial cryptography and data security. FC 2014. Lecture notes in computer science, vol 8438. Springer, Berlin, Heidelberg. https://doi.org/10.1007/978-3-662-44774-1_5
43. Gobbo N, Merlo A, Migliardi M (2013) A Denial of Service attack to GSM networks via attach procedure. In: Cuzzocrea A, Kittl C, Simos DE, Weippl E, Xu L (eds) Security engineering and intelligence informatics. CD-ARES 2013. Lecture notes in computer science, vol 8128. Springer, Berlin, Heidelberg. https://doi.org/10.1007/978-3-642-40588-4_25
44. Traynor P, Lin M, Ongtang M, Rao V, Jaeger T, McDaniel P, La Porta T (2009) On cellular botnets: measuring the impact of malicious devices on a cellular network core. In: Proceedings of the 16th ACM conference on computer and communications security. ACM, pp 223–234
45. Khan M, Ahmed A, Cheema AR (2008) Vulnerabilities of UMTS access domain security architecture. In: Ninth ACIS international conference on software engineering, artificial intelligence, networking, and parallel/distributed computing, SNPD 2008. IEEE, pp 350–355
46. Rajakumaran G, Venkataraman N, Mukkamala RR (2020) Denial of Service attack prediction using gradient descent algorithm. SN Comput Sci 1:45. https://doi.org/10.1007/s42979-019-0043-7
47. Zhang J, Chen C, Xiang Y, Zhou W, Vasilakos AV (2013) An effective network traffic classification method with unknown flow detection. IEEE Trans Netw Serv Manage 10(2)
48. Soja Rani S, Reeja SR (2020) A survey on different approaches for malware detection using machine learning techniques. In: Karrupusamy P, Chen J, Shi Y (eds) Sustainable communication networks and application. ICSCN 2019. Lecture notes on data engineering and communications technologies, vol 39. Springer, Cham. https://doi.org/10.1007/978-3-030-34515-0_42

49. Eskandari M, Khorshidpour Z, Hashemi S (2013) HDM-analyser: a hybrid analysis approach based on data mining techniques for malware detection. J Comput Virol Hacking Tech 9:77–93

50. Dali Z, Hao J, Ying Y, Wu D, Weiyi C (2017) DeepFlow: deep learning-based malware detection by mining Android application for abnormal usage of sensitive data. In: 2017 IEEE symposium on computers and communications (ISCC), pp 438–443

51. Ye Y, Chen L, Hou S, Hardy W, Li X (2017) DeepAM: a heterogeneous deep learning framework for intelligent malware detection. Knowl Inf Syst 54:265–285

52. Martín A, Menéndez HD, Camacho D (2017) MOCDroid: multi-objective evolutionary classifier for Android malware detection. Soft Comput 21:7405–7415

53. Painter N, Kadhiwala B (2018) Machine-learning-based android malware detection techniques—a comparative analysis. In: Mishra D, Nayak M, Joshi A (eds) Information and communication technology for sustainable development. Lecture notes in networks and systems, vol 9. Springer, Singapore. https://doi.org/10.1007/978-981-10-3932-4_19

54. Shijo P, Salim A (2015) Integrated static and dynamic analysis for malware detection. Proc Comput Sci 46:804–811

55. Yerima S, Sezer S, Muttik I (2014) Android malware detection using parallel machine learning classifiers. In: 8th international conference on next generation mobile applications, services and technology. IEEE, pp 37–42

56. Feizollah A, Anuar N, Salleh R, Wahab A (2015) A review on feature selection in mobile malware detection. Digit Investig 13:22–37

57. Xu Z, Wen C, Qin S, Ming Z (2018) Effective malware detection based on behaviour and data features. In: Qiu M (ed) Smart computing and communication. SmartCom 2017. Lecture notes in computer science, vol 10699. Springer, Cham. https://doi.org/10.1007/978-3-319-738 30-7_6

58. Ye Y, Chen L, Hou S et al (2017) DeepAM: a heterogeneous deep learning framework for intelligent malware detection. Knowl Inf Syst:1–21

59. Tamersoy A, Roundy K, Chau DH (2014) Guilt by association: large scale malware detection by mining file-relation graphs. In: ACM SIGKDD international conference on knowledge discovery and data mining

60. Badotra S, Singh J (2019) Creating firewall in transport layer and application layer using software defined networking. In: Saini H, Sayal R, Govardhan A, Buyya R (eds) Innovations in computer science and engineering. Lecture notes in networks and systems, vol 32. Springer, Singapore. https://doi.org/10.1007/978-981-10-8201-6_11

61. Heena JS (2016) Development of top layer web based filtering firewall using software defined networking. Int J Adv Res Comput Softw Eng 6(6)

62. Kaur K, Singh J (2016) Building stateful firewall over software defined networking. In: Information systems design and intelligent applications. Springer

63. Hu H, Han W, Ahn G-J, Zhao Z (2014) FLOWGUARD: building robust firewalls for software-defined networks. In Proceedings of the third workshop on hot topics in software defined networking. ACM, pp 97–102

64. Hu H, Han W, Ahn G-J, Zhao Z (2014) FLOWGUARD: building robust firewalls for software-defined networks. In: Proceedings of the third workshop on hot topics in software defined networking. ACM, pp 97–102

65. Likhar P, Shankar Yadav R (2020) Stealth firewall: invisible wall for network security. In: Saini H, Sayal R, Buyya R, Aliseri G (eds) Innovations in computer science and engineering. Lecture notes in networks and systems, vol 103. Springer, Singapore. https://doi.org/10.1007/978-981-15-2043-3_46

66. Ingham K, Forrest S (2002) A history and survey of network firewalls. ACM J:1–42

67. Chen S, Iyer R, Whisnant K (2002) Evaluating the security threat of firewall data corruption caused by transient instruction errors. In: International conference on dependable systems and network, Washington, DC, pp 495–504. https://doi.org/10.1109/DSN.2002.1028938

68. Maram B, Gnanasekar JM, Manogaran G et al (2019) Intelligent security algorithm for UNICODE data privacy and security in IoT. SOCA 13:3–15. https://doi.org/10.1007/s11761-018-0249-x
69. Suryavanshi H, Bansal P (2012) An improved cryptographic algorithm using UNICODE and universal colors. IEEE. 978-1-4673-1989-8/12/$31.00©2012
70. Dutta S, Das T, Jash S, Patra D, Paul P (2014) A cryptography algorithm using the operations of genetic algorithm and pseudo random sequence generating functions. Int J Adv Comput Sci Technol 3(5). ISSN 2320-2602
71. BalajeeMaram K, Gnanasekar JM (2015) Lightweight cryptographic algorithm to improve avalanche effect for data security using prime numbers and bit level operations. Int J ApplEng Res 10(21):41977–41983. ISSN 0973-4562
72. Kumar M (2018) Advanced RSA cryptographic algorithm for improving data security. In: Bokhari M, Agrawal N, Saini D (eds) Cyber security. Advances in intelligent systems and computing, vol 729. Springer, Singapore. https://doi.org/10.1007/978-981-10-8536-9_2
73. Nath A, Ghosh S, Malik MA Symmetric key cryptography using random key generator, vol 2, pp 239–244
74. Kaaniche N, Laurent M (2017) Data security and privacy preservation in cloud storage environments based on cryptographic mechanisms. Comput Commun 111:120–141
75. Kejariwal A, Bera S, Dash S, Maity S (2021) Information security using multimedia steganography. In: Mishra D, Buyya R, Mohapatra P, Patnaik S (eds) Intelligent and cloud computing. Smart innovation, systems and technologies, vol 153. Springer, Singapore. https://doi.org/10.1007/978-981-15-6202-0_65
76. Khari M, Garg AK, Gandomi AH, Gupta R, Patan R, Balusamy B (2020) Securing data in internet of things (IoT) using cryptography and steganography techniques. IEEE Trans Syst Man Cybern Syst 50(1):73–80. https://doi.org/10.1109/TSMC.2019.2903785
77. AlRoubiei M, AlYarubi T, Kumar B (2020) Critical analysis of cryptographic algorithms. In: 2020 8th international symposium on digital forensics and security (ISDFS), Beirut, Lebanon, pp 1–7. https://doi.org/10.1109/ISDFS49300.2020.9116213
78. Moayed MJ, Ghani AAA, Mahmod R (2008) A survey on cryptography algorithms in security of voting system approaches. In: 2008 international conference on computational sciences and its applications, Perugia, pp 190–200. https://doi.org/10.1109/ICCSA.2008.42
79. Chapple M, Stewart JM, Gibson D (2018) Managing identity and authentication. (ISC) 2 CISSP Certified information systems security professional official study guide. John Wiley & Sons
80. Identity as a Service (n.d.) Retrieved September 19, 2020, from https://resources.infosecinstitute.com/category/certifications-training/cissp/domains/identity-and-access-management/identity-as-a-service/
81. What is Kerberos and how does it work? (n.d.). Retrieved September 19, 2020, from https://web.mit.edu/kerberos/krb5-1.5/krb5-1.5.4/doc/krb5-install/What-is-Kerberos-and-How-Does-it-Work_003f.html
82. What is IDaaS? Understanding identity as a service and its applications (n.d.). Retrieved September 19, 2020, from https://www.okta.com/identity-101/idaas/
83. Implement and manage authorization mechanisms—CISSP exam (2020, July 16). Retrieved September 24, 2020, from https://www.itperfection.com/cissp/identity-and-access-management-domain-iam/implement-and-manage-authorization-mechanisms/
84. Tripwire Guest Authors Jan 29, 2 (2020, January 29) On authorization and implementation of access control models. Retrieved September 24, 2020, from https://www.tripwire.com/state-of-security/security-data-protection/authorization-implementation-access-control-models/
85. CISSP rapid review: access control (n.d.). Retrieved September 23, 2020, from https://www.microsoftpressstore.com/articles/article.aspx?p=2201319
86. Sullivan D (n.d.). Chapter 5. In: The definitive guide to security management

87. McMillan T, Abernathy RM (2018) Security assessment and testing. In: CISSP cert guide. Pearson, Indianapolis, Indiana
88. Disaster recovery plan (n.d.). Retrieved September 24, 2020, from https://www.ibm.com/services/business-continuity/disaster-recovery-plan
89. Asset Security (n.d.). Retrieved September 23, 2020, from https://resources.infosecinstitute.com/category/certifications-training/cissp/domains/asset-security/

Printed in the United States
by Baker & Taylor Publisher Services